The small hive beetle - a growing problem in the 21st century

Edited by Norman L. Carreck

I B R A

INTERNATIONAL BEE
RESEARCH ASSOCIATION

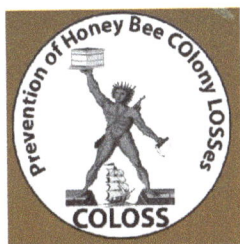

Prevention of Honey Bee COLony LOSSes
COLOSS

crea
Consiglio per la ricerca in agricoltura
e l'analisi dell'economia agraria

The small hive beetle - a growing problem in the 21st century

The small hive beetle - a growing problem in the 21st century

Edited by Norman L. Carreck

Layout by Valerie Rhenius
Preproduction by D&P Design and Print

ISBN: 978-0-86098-278-4

First edition July 2017. Jointly published by the International Bee Research Association, 91 Brinsea Road, Congresbury, Bristol, BS49 5JJ, UK and Northern Bee Books, Scout Bottom Farm, Mytholmroyd, Hebden Bridge, HX7 5JS, UK.

Orders: bookshop@ibra.org.uk or www.northernbeebooks.co.uk

The International Bee Research Association is a Company limited by Guarantee, Registered in England and Wales, Reg. No. 463819, Registered Office: 91 Brinsea Road, Congresbury, Bristol, BS49 5JJ, UK, and is a Registered Charity No. 209222.

http://www.ibra.org.uk/
https://www.facebook.com/IBRAssociation
https://twitter.com/ibra_bee

Notes:

1. Throughout this publication, the common name "small hive beetle" ("SHB") is used to mean the species *Aethina tumida* Murray 1867 (Coleoptera: Nitidulidae). Where other species of beetle are referred to, they are named in full.

2. All illustrations have been contributed by the first named author, except where otherwise noted.

3. The views expressed are not necessarily those of the International Bee Research Association.

Contents

ONE

Introduction - the development of our understanding of the biology of the small hive beetle

Norman L Carreck

The small hive beetle (SHB) *Aethina tumida* (Murray) is a relatively recent problem for beekeeping. Neither of the two authoritative texts of the bee pathologist, "Honey bee pathology" (Bailey and Ball, 1991) or "Honey bee pests predators and diseases" (Morse and Flottum, 1997) mention it at all. Although known for some time in bee hives within its natural sub-Saharan African range, it was only since its unexpected discovery as a pest in Florida, USA in June 1998 that it attracted much attention. Since then a number of key popular articles, reviews and primary research papers on the SHB have added greatly to our knowledge.

An early paper was that by Hood (2000), which described the beetle's arrival in the USA, probably in 1996. By 2000 it had spread to 12 US states, and caused considerable damage. A key paper by Elzen *et al.* (2001) compared the behaviour of SHB on European *Apis mellifera* in Florida, USA, with that on *A. m. capensis*, one of their native hosts, in Grahamstown, South Africa, their native range. Later that year, Neumann *et al* (2001) described how SHB could be reared under laboratory conditions. The following year, Mostafa and Williams (2002) described the incidence of the SHB in Egypt, out of its sub-Saharan

African range. In 2004, Hood wrote an authoritative review on the SHB, outlining its distribution to date, behaviour and economic importance. In 2006, Cabanillas and Elzen described the laboratory study of the effect of entomopathogenic nematodes on SHB, with the aim of using them for biological control.

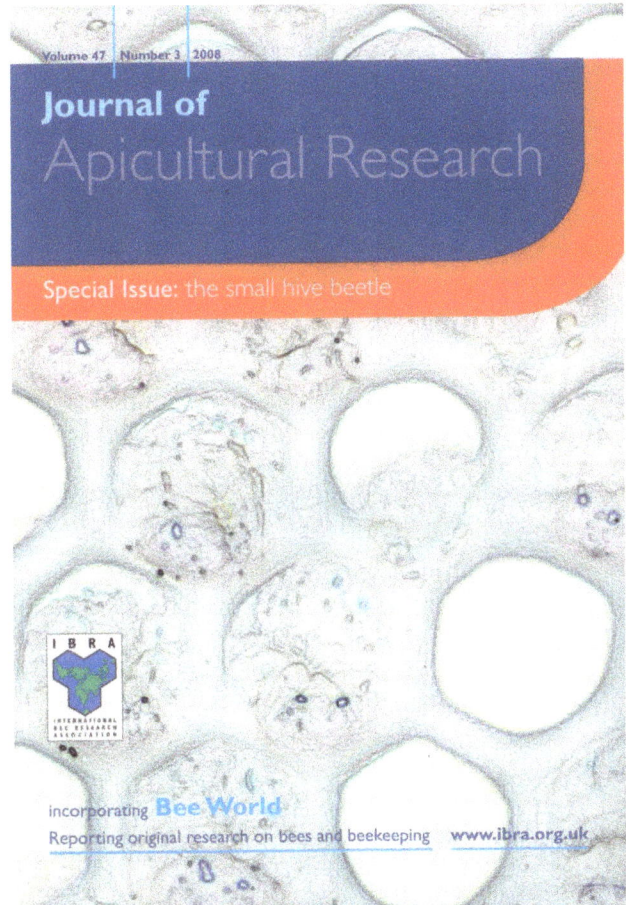

Fig. 1. The small hive beetle Special Issue of the *Journal of Apicultural Research*.

2

In 2008, Issue 47(3) of the *Journal of Apicultural Research* was entirely devoted to the SHB (Fig. 1). The editorial (Neumann and Ellis, 2008) reviewed the then current distribution of the pest and introduced the issues. One key paper by Bucholz *et al.* (2008) described laboratory experiments to test whether SHB could reproduce on a range of alternative food sources. They found that in the laboratory SHB could reproduce on fruit, but clearly preferred hive products, but it was unclear to what extent they may make use of these alternative food sources in the field. The paper by de Guzman *et al* (2008) showed that both commercial honey bees and the "Russian hygienic" bees in the USA were equally able to detect and uncap brood infested by eggs and larvae of SHB. Finally, Levot (2008) described a new design of insecticidal trap for SHB.

In 2012, two papers were published by Neumann *et al* on the long range dispersal of SHB, and by Spiewok and Neumann on the sex ratios of beetles in Australia and in South Africa. In the following year the COLOSS (Prevention of honey bee COLonly lOSSes) Association published its *BEEBOOK* (Williams *et al.*, 2012). This included a chapter on the SHB (Neumann *et al*, 2013) which covers the diagnosis and identification, laboratory rearing and marking, studies on biological control, and field techniques. This chapter has recently been translated into Spanish.

In June 2014, SHB was found in Lupon, Davao Oriental, Philippines, the first finding in Asia (Cervancia et al., 2016). The arrival of the SHB in Italy in September 2014 (Mutinelli, 2014; Mutinelli *et al.*, 2014; Palmeri *et al.*, 2015; Quigley,

Fig. 2. Issue 91(4) of *Bee World*.

2015; Chapter Two; Fig. 2), following a "scare" in 2004 in Portugal (Valério da Silva, 2014) seems to constitute its permanent arrival in Europe. In 2017, up to 1 June, five recent positive cases have been detected in southern Italy, including both wild and managed nests, and larvae and adult beetles.

Fig. 3. Participants at the COLOSS workshop. Photo: Noa Simon Delso.

In March 2015, the SHB was discovered in Piracicaba, São Paulo State, Brazil, the first finding on the continent of South America (Al Toufailia *et al.*, 2017). Further spread to other countries seems probable (Neumann *et al.*, 2016), although it is difficult to predict its impact on bees under particular local conditions. There has inevitably been, therefore, a resurgence of interest in the SHB by both researchers and beekeepers.

In response to this, a meeting entitled "A European Strategy for small hive beetles *Aethina tumida*" was held in Bologna, Italy on 19 and 20 February 2015. This was organised by the bee research unit of the Consiglio per la ricerca in agricoltura e l'analisi dell'economia agraria, (CREA-API) under the auspices of the newly formed COLOSS Small Hive Beetle Task Force. The first day was planned as a COLOSS Workshop, and was attended by 52

people from 17 countries, including most of the people with active research or other experience of SHB (Fig. 3). The meeting produced a number of important conclusions and recommendations (See Appendix). The second day was an Extension Day attended by about 120 people, mostly beekeepers, veterinarians and officials from the Italian Ministries of Agriculture and of Health (Fig. 4).

This book is intended for beekeepers and others interested in the beetle, and is mainly derived from the talks presented over those two days in Bologna. In Chapter Two, Franco Mutinelli and colleagues describe the circumstances of the original discovery of SHB in Italy and the initial reaction measures. In Chapter Three, Marie-Pierre Chauzat and colleagues set this in the context of EU and international legislation. Then, in Chapter Four, Peter Neumann outlines the possible implications of the discovery and of the future spread of SHB given our knowledge of the beetle and its biology. Next, in Chapter Five, Jeff S Pettis describes the effects of the beetle in the USA, the region where damage has been most severe. In Chapter Six, Christian Pirk and colleagues describe the effects of the beetle in its native range, Sub-Saharan Africa on one of its its native honey bee hosts *Apis mellifera scutellata*. Finally in Chapter Seven, Robert Spooner-Hart and colleagues describe the effects of the SHB in Australia, on both the imported honey bee and on native stingless bees.

I hope that the combined knowledge of these scientists with practical experience of the beetle will enable beekeepers to effectively control it wherever it is found.

Fig. 4. Participants at the COLOSS extension day. Photo: Norman L Carreck.

References

Al Toufailia, H., Alves, D. A., Bená, D. C., Bento, J. M. S., Iwanicki, N. S. A., Cline, A. R., Ellis, J. D. & Ratnieks, F. L. W. (2017) First record of small hive beetle, *Aethina tumida* Murray, in *South America. Journal of Apicultural Research*, 56(1), 76-80.
http://dx.doi.org/10.1080/00218839.2017.1284476

Bailey, L. & Ball, B. V. (1991). *Honey bee pathology.* Academic Press; London, UK. 193 pp. ISBN: 0-12-073481-8

Buchholz, S. B., Schäfer, M. O., Spiewok, S., Pettis, J. S., Duncan, M., Ritter, W., Spooner-Hart, R. & Neumann, P. (2008). Alternative food sources of *Aethina tumida* (Coleoptera: Nitidulidae). *Journal of Apicultural Research*, 47(3), 202-209.
http://dx.doi.org/10.3896/IBRA.1.47.3.08

Cabanillas, H. E. & Elzen, P. J. (2006). Infectivity of entomopathogenic nematodes (Steinernematidae and Heterorhabditidae) against the small hive beetle *Aethina tumida* (Coleoptera: Nitidulidae). *Journal of Apicultural Research*, 45(1), 49-50.
http://dx.doi.org/10.1080/00218839.2006.11101314

Cervancia, C. R., de Guzman, L. I., Polintan,, E. A., Dupo, A. L. B. & Locsin, A. A. (2016) Current status of small hive beetle infestation in the Philippines. *Journal of Apicultural Research*, 55(1), 74-77.
http://dx.doi.org/10.1080/00218839.2016.1194053

de Guzman, L. I., Frake, A. & Rinderer, T. E. (2008). Detection and removal of brood infested with eggs and larvae of small hive beetles (*Aethina tumida* Murray) by Russian honey bees. *Journal of Apicultural Research*, 47(3), 216-221.
http://dx.doi.org/10.3896/IBRA.1.47.3.10

Elzen, P. J., Baxter, J. R., Neumann, P., Solbrig, A., Pirk, C. W. W., Hepburn, H. R., Westervelt, D. & Randall, C. (2001) Behaviour of African and European sub-species of *Apis mellifera* towards the small hive beetle *Aethina tumida. Journal of Apicultural Research*, 40(1), 40-41.
http://dx.doi.org/10.1080/00218839.2001.11101049

Hood, M. W. M. (2000). Overview of the small hive beetle, *Aethina tumida*, in North America. *Bee World*, 81(3), 129-137.
http://dx.doi.org/10.1080/0005772X.2000.11099483

Hood, M. W. M. (2004). The small hive beetle, *Aethina tumida*: a review. *Bee World*, 85(3), 51-59.
http://dx.doi.org/10.1080/0005772X.2004.11099624

IZSVe (2017).

http://www.izsvenezie.com/aethina-tumida-in-italy/

Levot, G. W. (2008). An insecticidal refuge trap to control adult small hive beetle, *Aethina tumida* Murray (Coleoptera: Nitidulidae) in honey bee colonies. *Journal of Apicultural Research,* 47(3), 222-228.

http://dx.doi.org/10.3896/IBRA.1.47.3.11

Morse, R. A. & Flottum, K. (Eds) (1997). *Honey bee pests, predators and diseases (3rd Ed.).* A. I. Root Co.; Medina, Ohio, USA. 718 pp. ISBN: 0-936028-10-6

Mostafa, A. M. & Williams, R. N. (2002). New record of the small hive beetle in Egypt and notes on its distribution and control. *Bee World,* 83(3), 99-108.

http://dx.doi.org/10.1080/0005772X.2002.11099549

Mutinelli, F. (2014). The 2014 outbreak of the small hive beetle in Italy. *Bee World,* 91(4), 88-89.

http://dx.doi.org/10.1080/0005772X.2014.11417618

Mutinelli, F., Montarsi, F., Federico, G., Granato, A., Maroni Ponti, A., Grandinetti, G., Ferrè, N., Franco, S., Duquesne, V., Rivière, M.-P., Thiéry, R., Henrikx, P., Ribière-Chabert, M. & Chauzat, M.-P. (2014). Detection of *Aethina tumida* Murray (*Coleoptera: Nitidulidae.*) in Italy: outbreaks and early reaction measures. *Journal of Apicultural Research,* 53, 569-575.

http://dx.doi.org/10.3896/IBRA.1.53.5.08

Neumann, P. & Ellis, J. D. (2008). The small hive beetle (*Aethina tumida* Murray, Coleoptera: Nitidulidae): distribution, biology and control of an invasive species. *Journal of Apicultural Research,* 47(3), 181-183.

http://dx.doi.org/10.3896/IBRA.1.47.3.01

Neumann, P., Evans, J. D., Pettis, J. S., Pirk, C. W. W., Schäfer, M. O., Tanner, G. & Ellis, J. D. (2013). Standard methods for small hive beetle research. In *V. Dietemann, J. D. Ellis & P. Neumann (Eds) The COLOSS BEEBOOK: Volume II: Standard methods for* Apis mellifera *pest and pathogen research. Journal of Apicultural Research,* 52(4),

http://dx.doi.org/10.3896/IBRA.1.52.4.19

Neumann, P., Hoffmann, D., Duncan, M., Spooner-Hart, R. & Pettis, J. S. (2012). Long-range dispersal of small hive beetles. *Journal of Apicultural Research,* 51(2), 214-215.

http://dx.doi.org/10.3896/IBRA.1.51.2.11

Neumann. P., Pettis, J. S., Schäfer, M. O. (2016) *Quo vadis Aethina tumida?* Biology and control of small hive beetles. *Apidologie,* 47(3), 427-466.

http://dx.doi.org/10.1007/s13592-016-0426-x

Neumann, P., Pirk C. W. W., Hepburn, H. R., Elzen, P. J. & Baxter, J. R. (2001). Laboratory rearing of small hive beetles *Aethina tumida* (Coleoptera, Nitidulidae). *Journal of Apicultural Research,* 40(3-4), 111-112.

http://dx.doi.org/10.1080/00218839.2001.11101059

Palmeri, V., Scirtò, G., Malacrinò, A., Laudani, F. & Campolo, O. (2015). A new pest for European honey bees: first report of *Aethina tumida* Murray (Coleoptera Nitidulidae) in Europe. *Apidologie,* 46(4), 527-529.

http://dx.doi.org/10.1007/s13592-014-0343-9

Quigley, A. S. (2015) The nightmare returns for Calabrian beekeepers. One year on and SHB is back to haunt them! *Bee World,* 92(2) 42.

http://dx.doi.org/10.1080/0005772X.2015.1118967

Spiewok, S. & Neumann, P. (2012). Sex ratio and dispersal of small hive beetles. *Journal of Apicultural Research*, 51 (2), 216-217. http://dx.doi.org/10.3896/IBRA.1.51.2.12

Williams, G. R., Dietemann, V., Ellis, J. D. & Neumann, P. (2012). An update on the COLOSS network and the "BEEBOOK: standard methodologies for *Apis mellifera* research". *Journal of Apicultural Research*, 51(2), 151-153. http://dx.doi.org/10.3896/IBRA.1.51.2.01

Valério da Silva, M. J. (2014). The first report of *Aethina tumida* in the European Union, Portugal, 2004. *Bee World*, 91(4), 90 -91. http://dx.doi.org/10.1080/0005772X.2014.11417619

Norman L Carreck[1,2]

[1]International Bee Research Association, 91 Brinsea Road, Congresbury, Bristol, BS49 5JJ, UK.
Email: carrecknl@ibra.org.uk
[2]Laboratory of Apiculture and Social Insects, University of Sussex, Falmer, Brighton, East Sussex, BN1 9QG, UK.

TWO

The small hive beetle in Italy

Franco Mutinelli, Giovanni Federico, Fabrizio Montarsi, Anna Granato, Claudia Casarotto, Gianluca Grandinetti, Marie-Pierre Chauzat and Andrea Maroni Ponti

Introduction

The first detection of the small hive beetle in Italy was made on 5 September 2014 (Palmeri et al., 2015). Three nucleus hives containing honey bees and located in a clementine (citrus) orchard in the Calabria region (southern Italy) at the locality of Sovereto (N 38.45474 E 15.94110), Gioia Tauro municipality (Fig. 1) were found to be heavily infested with adult and larval beetles. Upon this discovery, the three bee nuclei were wrapped and brought back to the Università "Mediterranea" at Reggio Calabria, where they were killed and immediately deep frozen. A sample of approximately 15 adult and 15 larval beetles were taken for identification. The specimens were identified as Aethina tumida based on morphological characteristics. On 10 September 2014, the specimens were sent to the Italian National Reference Laboratory (NRL) (Istituto Zooprofilattico Sperimentale delle Venezie). The species was confirmed through morphological identification. Some adults and larvae were sent to the European Reference Laboratory (EU RL) in Sophia-Antipolis (France) where the species was confirmed subsequently through morphological identification on 15 September and through molecular techniques on 17 September 2014. On 18 September, A. tumida detection in Italy was reported to the OIE (World Organisation for Animal Health) (Mutinelli, 2014; Mutinelli et al., 2014).

The site where the beetles were discovered was treated with chlorpyriphos methyl on the 5 September by the University team, and on 11 September by the veterinary services and the University team. The pesticide was mixed with water and poured directly on the soil at 20 l per 15 m² of soil surface. A third insecticide treatment was applied on the site, by abundantly spraying a 1% solution of cypermethrin and tetramethrin (6.85% and 1.25% concentration respectively) after having ploughed the soil on 17 September. Two new bee nuclei were installed on 17 September in close vicinity (50 m) of the initial nuclei site and were fitted with traps (E H Thorne (Beehives) Ltd; Rand, UK) (Schäfer et al., 2008; 2010). The traps were reduced in size by two thirds so that they could fit through the nucleus hive entrances. Adult beetles were found in the nuclei on 10 October, so the nuclei were destroyed on 14 October. The soil under and around the hives was abundantly sprayed with a 1% solution of cypermethrin and tetramethrin by the veterinary service team.

During October 2014, the three infested and frozen nuclei were thoroughly investigated at the Italian NRL in Legnaro (Padova) revealing the presence of 19 larvae; 12

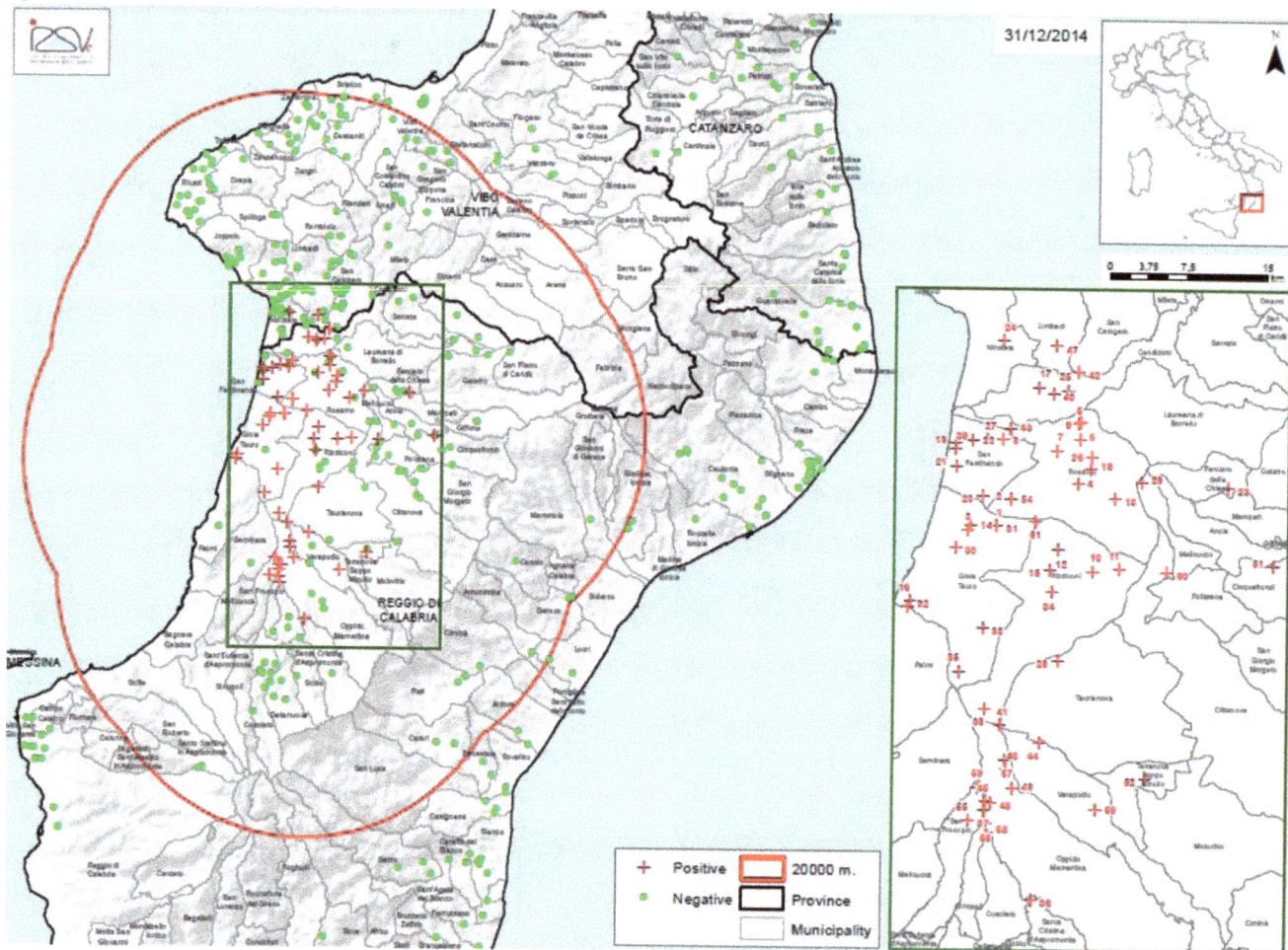

Fig. 1. Location of apiaries infested with SHB in the Calabria region (at 31 December 2014). Red cross: infested apiary. Green dot: visited apiary but no SHB detected. The control zone (20 km from infested apiaries) is delineated by the red circle.

recovered from adult bees and 7 from combs. Nucleus No. 1 contained 4 larvae, nucleus No. 2, 12 and nucleus No. 3, 3 larvae. No adult beetles were found.

Field observations

On 17 September, we (F.M., G.F. and M.P.C.) dug the soil at the initial discovery site to search for SHB pupae (Fig 2.). Five dipteran pupae and one larvae of a big coleopteran

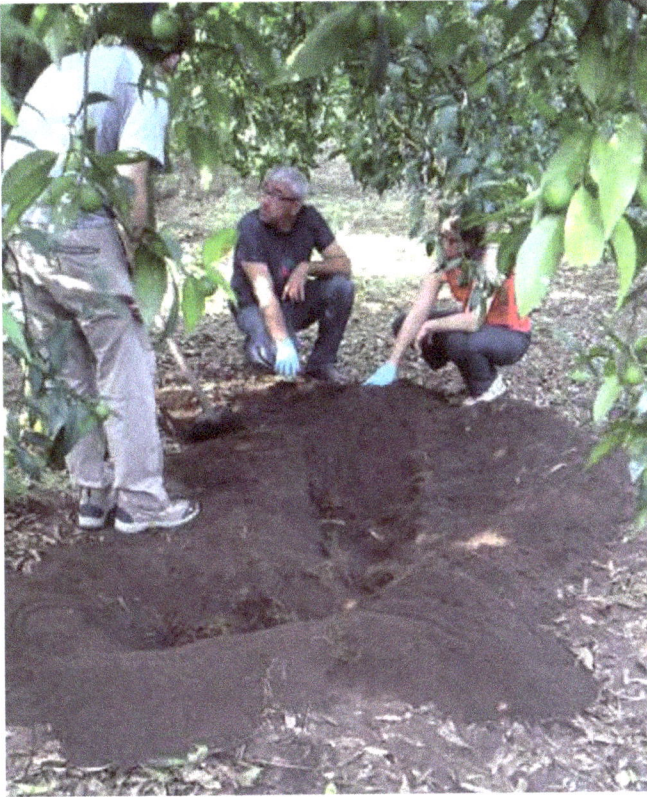

Fig. 2. Digging the soil at the original site of discovery.

(according to the size and the morphological characteristics it was not *A. tumida* but probably a xylophagus species) were found. The soil was then ploughed and treated with a 1% solution of cypermethrin and tetramethrin. On 18 September, traps established in the two newly installed nuclei were checked. No beetles were found. One nucleus was inspected visually and no beetles were found.

From 16 to 17 September 2014, a team composed of beekeepers, biologists and official veterinarians, including three authors of this paper (F.M., G.F. and M.P.C.), visited five apiaries, all located in the vicinity of the original discovery site. A proportion of colonies, ranging from 20 to 50%, in each apiary were inspected for SHB. Hive examination may provide an early diagnosis of infestation. This procedure is accomplished with two people, usually the beekeeper and the official veterinarian, one to work the colony and the second to collect the beetles. They proceed as follows: remove the lid from the colony and examine the inner part of the lid for the presence of adult beetles. Place it at the side of the bee hive. Lightly smoke the colony, remove the outermost frame in the super and/ or in the hive (Dadant-Blatt type), and quickly examine both faces of the comb. The outermost frame is then placed at the side of the hive and all the other frames undergo the same visual inspection one by one. Once inspected, each frame is reintroduced in the super or in the hive in the same order using the room left available by the outermost frame in order to prevent robbing. Then the inside faces of the hive and the bottom board are carefully examined. When all frames have been inspected, they are replaced in the original position as is the outermost frame and the hive is closed. When present, the frames of both the super and the hive should be thoroughly examined. This procedure allowed us to detect both beetle adults and larvae. In each apiary, 5 to 27 traps were installed during the visit.

Adult SHB were detected on 17 September 2014 through visual observation in one apiary located 2 km from the original discovery site in Collina locality, Rosarno municipality, Reggio Calabria province. Six colonies were examined out of 41 located in this apiary. SHB were observed in four colonies. In total, 7 adults were collected. A set of 27 colonies were equipped with traps. On 18 September, all 41 colonies were inspected thoroughly. Eighteen adult beetles were collected from 12 hives. About six adult beetles escaped the sampling by flying away. Of the 27 traps, two hosted adult small hive beetles. No brood destruction was observed in any of the 41 colonies inspected. Out of the four positive colonies detected the previous day, two were positive again the next day. The entire apiary was destroyed according to the rule in force.

Two weak nucleus bee colonies were placed in the same site mentioned above on 4 November and they were found positive for SHB adults on 13 November 2014. Both nuclei were destroyed on 14 November 2014.

European and national regulatory framework

A. tumida was exotic to Europe until September 2014. Being a notifiable disease in the European Union (European Commission, 1982; 1992; 2004) and an OIE listed disease (OIE, 2014), any identification of the pest must be reported to national competent authorities, to the European Commission and to the OIE. Member states of the European Union have to implement passive surveillance programmes specifically directed to this species. See Chapter 3 for more details. An input to this activity comes also from the recent EU epidemiological study on honey bee colony losses (Chauzat *et al.*, 2014). In case of detection, contaminated apiaries should be destroyed. In the present case, traps were installed in all the colonies located in apiaries with no signs of infestation (meaning presence of SHB adults, larvae or damaged frames).

The importation of honey bees is strictly regulated in the European Union. Only queen bees with attendants (20 maximum) should be imported in Europe according to a stringent authorization pathway (European Commission, 2010). Packages should be controlled at the place of origin and at destination. Furthermore, queen bees should come from an area of at least 100 km radius that is not subject to any restrictions associated with the occurrence of SHB and where this infestation is absent (Mutinelli, 2011). In 2004, the controls carried out on queens and accompanying workers of *Apis mellifera ligustica* legally imported into the region of Alentejo, Portugal from Texas, USA prevented a potential introduction of *A. tumida* (Murilhas, 2004; Neumann & Ellis, 2008; Valério da Silva, 2014). The queen bees, attendants, food and packaging as well as the apiaries where the queens went to destination before the end of the check on the entire imported batch, were destroyed.

Following the confirmed occurrences of the small hive beetle in Italy, an implementing decision concerning certain protective measures was published on 16 December 2014 (European Commission, 2014). Italy shall ensure the implementation of the ban on the dispatch of consignments of: i. honey bees; ii. bumble bees; iii. unprocessed apiculture by-products; iv. beekeeping equipment; and v.

comb honey intended for human consumption from the areas listed in the Annex (the whole territory of Calabria and Sicily regions) to other areas of the Union as well as the carrying out of immediate inspections and epidemiological investigations in the infested area. This Decision was due to expire on 31 May 2015 but was then extended until 30 November 2015 (European Commission, 2015).

Furthermore, the Commission asked the European Food Safety Authority (EFSA) to provide scientific and technical assistance concerning: 1. the currently employed diagnostic methods for the detection of SHB and the risk mitigation measures applied worldwide in relation to SHB in apiaries and in controlled establishments producing queens, as well as measures applied to domestic movements of colonies, queens and other honey bee products and by-products; 2. the best practices or strategies to be applied in an infected area in order to respectively eradicate or control the spread of the SHB. The EFSA scientific report on small hive beetle diagnosis and risk management options was published on 17 March 2015 (EFSA, 2015). An EFSA scientific opinion is expected by the end of 2015.

The EU RL for bee health in cooperation with the NRLs of Germany, UK and Italy has produced a leaflet to inform veterinary services, practitioners and beekeepers on the threat posed by SHB to honey bees in order to better identify any outbreak. This leaflet has been translated from English into other European languages and disseminated throughout the European Union and additional countries (European Union Reference Laboratory, 2015).

Following the detection of SHB in Italy, a restriction zone with a 20 km radius from the initial infested site was established (Fig. 1). Any movement of colonies or beekeeping material within this zone and between this zone and other areas was forbidden. In this zone, all apiaries should be visited and a proportion of hives fully inspected. This proportion was deemed appropriate to detect SHB presence at an expected prevalence of 5% with a 95% confidence interval (CI). Therefore, the number of inspected colonies varied according to the number of colonies in the apiary. A surveillance zone within a 100 km radius from the initial infested site was established (Fig. 3). Apiaries in this zone were sampled according to two strategies: the presence of risk factors or, if none, a random selection of apiaries within the zone. This allowed us to detect SHB at a 2% apiary level expected prevalence with a 95% CI (i.e. at least 149 apiaries randomly selected). The traps (E H Thorne (Beehives) Ltd.; Rand, UK; homemade corrugated plastic sheet; and Beetle Blaster® (Mann Lake; MN, USA) were installed in all hives under surveillance. Figs 1 and 3 show the results of inspections as at 31 December 2014. SHB has been detected in 59 apiaries and one natural honey bee colony in the Calabria region and in eighteen municipalities belonging to two provinces (province with number of infested apiaries in parentheses): Reggio Calabria (56 and a natural colony) and Vibo Valentia (3). This represents an area over 400 km^2. In

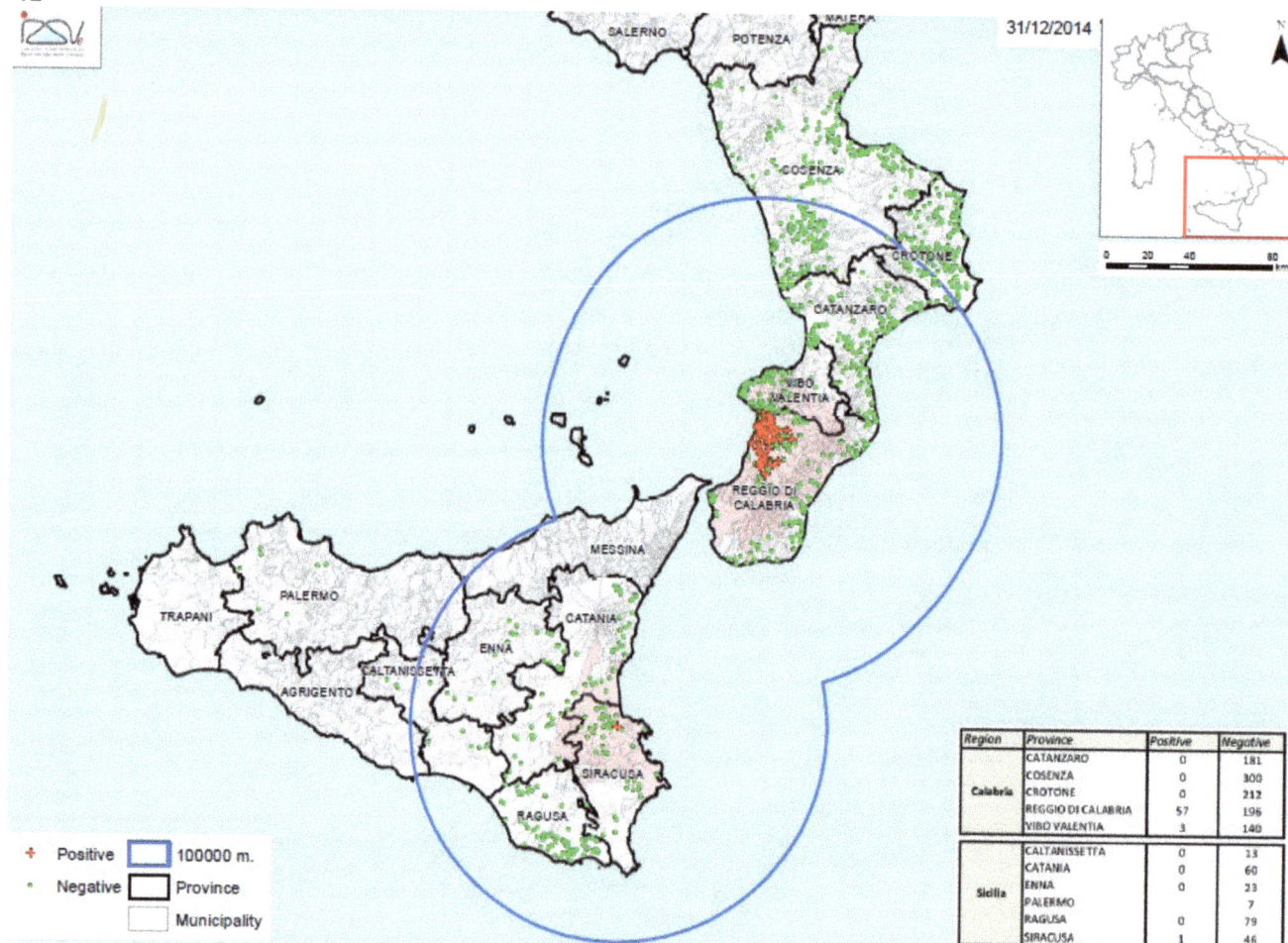

Fig. 3. Surveillance zone (100 km from the initial outbreaks in Calabria and Sicily) for SHB in Italy (31 December 2014). Red cross: infested apiary. Green dot: visited apiary but no SHB detected.

most cases, with the exception of six sites, three in Gioia Tauro, one respectively in Rizziconi, Candidoni and Cittanova municipalities where larvae have been detected (Fig. 1), only adult SHB were observed. In only one case, a single pupa was found (Gioia Tauro municipality) in the soil of an infested apiary that also contained adult and larval beetles. A single adult beetle was found in a natural honey bee colony (Gioia Tauro municipality). The colony and the combs were collected and the bees killed. Neither adult nor larval SHB were detected after a thoroughly investigation of bees and combs.

All the infested apiaries have been destroyed by the veterinary service team according to the decision taken by the Ministry of Health and shared with the National Beekeeping Associations during the meeting held in Lamezia Terme (Calabria) on 22 September 2014 and further confirmed in the meeting of the National Crisis Unit held in Rome on 11 December 2014. A Decree of the Ministry of Health of 19 November 2014 guaranteed compensation for colonies and beekeeping equipment destroyed due to infestation by A. tumida according to Italian Law 218/1988 (http://www.trovanorme.salute.gov.it/norme/dettaglioAtto?id=50871).

Apiary "stamping-out" was carried out according to a common protocol based on closure of the hives in the evening, killing the honey bees using sulphur dioxide spray, and burning the dead colonies on site. A soil treatment was abundantly applied in the apiary as 1% solution of cypermethrin and tetramethrin of a commercially available product after soil ploughing.

The territory coverage where the infested apiaries were located is 38.98% fruit trees and berry plantations, 25.42% olive groves, 20.34% annual crops associated with permanent crops, 5.08% non-irrigated arable land, 3.40% continuous urban fabric and complex cultivation patterns, 1.69% construction sites and principally occupied by agriculture, with significant areas of natural vegetation (Corine, 2006). The soil in the area where infested apiaries were detected is sandy, soft, warm and dry (Province di Reggio Calabria e di Vibo Valentia, 2013). The area is also much ventilated due to the vicinity to the sea.

For the 60 infested sites in Calabria, several other apiaries were visited, without reporting the presence of any SHB as of 7 July 2015: 222 apiaries within the protection zone (20 km from the infested sites); 452 apiaries within the surveillance zone (100 km); and 356 apiaries outside the two zones, for a total of 1,030 apiaries in 2014. From January to July 2015, 782 apiaries have been visited in the all territory of Calabria region.

On 7 November 2014 SHB were was detected in a 56 hive migratory apiary located in the municipality of Melilli, Siracusa province, Sicily region (Fig. 3). The epidemiological investigation carried out by the local veterinary service demonstrated that these hives had been in the area of the first detection of SHB (Gioia Tauro, Calabria region) from April to August 2014. As of 7 July 2015, no further detection of A. tumida in Sicily region has been reported according to the inspection carried out in 15 apiaries within the protection zone (10 km from the infested sites), 202 within the surveillance zone (100 km) and 11 outside the two zones for a total of 228 apiaries (Fig. 3) in 2014. Between January and July 2015, 318 apiaries have been visited in the all territory of Sicily region.

The results of field investigation are available on the websites of the Italian National Reference Laboratory:

http://www.izsvenezie.com/aethina-tumida-in-italy/

and the EU Reference Laboratory: https://sites.anses.fr/en/minisite/abeilles/eurl-bee-health-home

A preliminary field investigation aiming at detection of SHB on rotten fruits in Calabria citrus and kiwi orchards was carried out between December 2014 and January 2015. Among collected samples, seven different non-SHB Nitidulid species were identified on rotten citrus, while no coleopters were detected on rotten kiwi. Furthermore, no SHB were detected on rotten citrus (Mutinelli *et al.*, in press).

Hypothesis on the introduction of SHB into Italy

As mentioned in the risk assessment performed by EFSA, there are several possible scenarios for the introduction of SHB into Europe (EFSA, 2013). The port of Gioia Tauro is located near the initial infested apiary. About 2.5 million containers arrive each year in the 7.5 km long harbour. Being a transition platform, most of the containers undergo documentary and seal check, and then are transferred from the mother ship to daughter ships and sent to their final destination. Before transportation to other smaller ships or on trucks for their final destination, identity and physical controls are performed on a proportion of goods according to the EU Commission indications (custom inspections, plant health inspection, animal health inspection, i.e. Border Inspection Post (BIP), etc.). The list of port of origin/loading of the containers that arrived in the port of Gioia Tauro between January and August 2014 was requested to the competent authority. The port of Gioia Tauro (Reggio Calabria) is not authorized for live animal introduction.

No honey bee swarm had been noticed in the harbour before the notification of the presence of SHB in Italy, although a wasp's nest had been observed on one occasion two years before the event.

The illegal importation of honey bees or bee products has also been considered. This much-trafficked area of southern Italy is subject to many movements of honey bee colonies as well as bee equipment. In the flat territory of Reggio Calabria province, the number of colonies usually doubles each year starting from April for the blooming of citrus, from 10,000 to 20,000 colonies. Extra colonies come mainly from Sicily and from other parts of Italy to forage on the available plant resources, including citrus, eucalyptus, and chestnut. After the honey flow, colonies are returned to their place of origin, thus facilitating the possible spread of SHB. This area is a significant source of honey bees and queen bees for Italy, much of Europe, and for beekeepers on other continents, making honey production almost a secondary activity. From this point of view, the possible impact of SHB on the southern Italian beekeeping industry and honey bee production is devastating. Following the early reaction measures adopted by the Ministry of Health, regional veterinary service personnel were requested to investigate any risk movement of honey bees (i.e. migratory beekeeping during 2014), live bee materials (i.e. queen bees and package bees), and bee equipment (Order 0018842-P-12/09/2014 and 0020069-01/10/2014-DGSAF-COD_UO-P). Similar activity is in progress also in the other European countries. In addition to the measures adopted for the Calabria region (Regional Order No. 94 of 19/09/2014), a national surveillance programme for the detection of *A. tumida* was defined and further revised according to the EU Guidelines for the surveillance of the small hive beetle (*Aethina tumida*) infestation (Chauzat *et al.*, 2015). See Chapter Three for more details.

As the soil in this area is sandy, warm and dry (Province di Reggio Calabria e Vibo Valentia, 2013) and the area much ventilated due to the vicinity to the sea, it can be hypothesized that the conditions are not very favourable for *A. tumida*. It could therefore have been present long before its detection, so the risk of accidental movement elsewhere is much greater. This has been confirmed by the detection of *A. tumida* in hives in the migratory apiary in Sicily that had been in Gioia Tauro (Calabria region) from April to August 2014.

According to this consideration and risk analysis, regional veterinary services of all Italian regions are proceeding with apiary and colony inspections. Rapid actions are required to understand as soon as possible, how far SHB has spread from its initial detection site. This information is needed in order to define the strategy for the possible eradication of the pest (the one presently adopted), to adjust this strategy if necessary or, in the worst case scenario (where SHB becomes established as an endemic pest), to replace the eradication protocol with one based on control.

Conclusions

From this report, it is worth noting that visual inspection of the colonies proved to be much more sensitive than the use of traps. Detection of SHB was highly variable depending on the method (visual detection or traps), probably due to the low prevalence of adult beetles in the hives at the time of observation. It is recommended to leave the traps in the hive for at least 48 hours before examination (Neumann *et al.*, 2013). However, until further data are available, we recommend performing a thorough visual inspection of colonies to detect the presence of SHB, using traps as a useful complimentary tool, as traps cannot replace visual inspection and have rarely detected SHB.

The early reaction measures adopted the Italian Ministry of Health aimed at containment and possibly eradication of SHB in the Calabria and Sicily regions. According to the data provided by the active surveillance carried out in the infested regions, infestation seems to be confined to a limited area of Calabria region (protection zone) and a single outbreak in Sicily region. Unfortunately, the winter season prevented the continuation of the intense surveillance activity carried out from September to December 2014. However, the resumption of the surveillance activity in the infested area as well as in the rest of Italy, has not so far revealed any other outbreaks. Appropriate evaluation and decisions about the current strategy applied will follow accordingly.

Acknowledgments

The authors wish to thank the beekeepers and the regional and local Veterinary Services for their cooperation. Mr Francesco Artese is thanked for his professional and passionate cooperation with the veterinary services and beekeepers during field investigations, Mrs Monica Lorenzetto and Luciana Barzon for their contribution to data management and map preparation.

References

Chauzat, M.-P., Laurent, M., Riviere, M-P., Saugeon, C., Hendrikx, P., & Ribiere-Chabert, M. (2014). *A pan-European epidemiological study on honey bee colony losses 2012-2013*. European Union Reference Laboratory for honey bee health (EURL); Sophia-Antipolis, France. http://ec.europa.eu/food/animals/live_animals/bees/docs/bee-report_en.pdf

Chauzat, M.-P., Laurent, M., Brown, M., Kryger, P., Mutinelli, F., Roelandt, S. & Hendrikx, P. (2015). *Guidelines for the surveillance of the small hive beetle* (Aethina tumida) *infestation*. European Union Reference Laboratory for honey bee health (EURL); Sophia-   Antipolis, France. 19 pp. https://sites.anses.fr/en/minisite/abeilles/eurl-bee-health-home

Corine Land Cover. (2006). *(CLC2006) 100 m - version 12/2009*. http://www.eea.europa.eu/data-and-maps/data/corine-land-cover-2006-clc2006-100-m-version-12-2009.

Cuthbertson, A. G. S., Wakefield, M. E.; Powell, M. E., Marris, G., Anderson, H., Budge, G. E. & Brown, M. A. (2013). The small hive beetle *Aethina tumida*: A review of its biology and control measures. *Current Zoology*, 59, 644–653. http://www.currentzoology.org/paperdetail.asp?id=12275

European Commission. (1982). Council Directive 82/894/EEC of 21 December 1982 on the notification of animal diseases within the Community. *Official Journal of the European Union*, L378, 31 December 1982, pp. 58–62.

European Commission. (1992). Council Directive 92/65/EEC of 13 July 1992 laying down animal health requirements governing trade in and imports into the Community of animals, semen, ova and embryos not subject to animal health requirements laid down in specific Community rules referred to in Annex A (I) to Directive 90/425/EEC. *Official Journal of the European Union*, L268: 14 September 1992, pp. 54-72. [as amended by Commission Regulation (EC) No. 1398/2003 of 5 August 2003 amending Annex A to Council Directive 92/65/EEC to include the small hive beetle (*Aethina tumida*), the Tropilaelaps mite (*Tropilaelaps* spp.), Ebola and monkey pox. *Official Journal of the European Union*, L198: 6 August 2003, p. 3].

European Commission. (2004). Commission Decision of 1 March 2004 amending Council Directive 82/894/EEC on the notification of animal diseases within the Community to include certain equine diseases and certain diseases of bees to the list of notifiable diseases. (2004/216/EC). *Official Journal of the European Union*, L67, 5 March 2004, pp. 27-30.

European Commission. (2010). Commission Regulation (EU) No. 206/2010 of 12 March 2010 laying down lists of third countries, territories or parts thereof authorised for the introduction into the European Union of certain animals and fresh meat and the veterinary certification requirements. *Official Journal of the European Union*, L73, 20 March 2010, 120 pp.

European Commission. (2014). Commission Implementing Decision of 12 December 2014 concerning certain protective measures with regard to confirmed occurrences of the small hive beetle in Italy. (2014/909/EU). *Official Journal of the European Union*, L359, 16 December 2014, pp. 161-163.

European Commission. (2015). Commission Implementing Decision of 28 May 2015 amending Implementing Decision 2014/909/EU by extending the period of application of the protection measures in relation to the small hive beetle in Italy. (2015/838/EU). *Official Journal of the European Union*, L132, 29 May 2015, pp. 86-87.

European Food Safety Authority. (2013). Scientific opinion on the risk of entry of *Aethina tumida* and *Tropilaelaps* spp. in the EU. *EFSA Journal,* 11, 3128. http://dx.doi.org/doi:10.2903/j.efsa.2013.3128

European Food Safety Authority. (2015). EFSA scientific report on small hive beetle diagnosis and risk reduction options. *EFSA Journal*, 13, 4048. http://dx.doi.org/doi:10.2903/j.efsa.2015.4048

European Union Reference Laboratory. (2015) *Small hive beetle*. ANSES; France. 2 pp. https://sites.anses.fr/en/system/files/SHB_For_beekeepers_2015feb_0.pdf

Murilhas, A. M. (2004). *Aethina tumida* arrives in Portugal. Will it be eradicated? *EurBee Newsletter*, 2, 7-9.

Mutinelli, F. (2011). The spread of pathogens through trade in honey bees and their products (including queen bees and semen): overview and recent developments. *Revue Scientifique et Technique de l'Office International des Epizooties*, 30, 257-271. http://web.oie.int/boutique/index.php?page=ficprod&id_prec=945&id_produit=1062&lang=en&fichrech=1&PHPSESSID=bdb97a30dd3f9558ddde00714cbb6356

Mutinelli, F. (2014). The 2014 outbreak of the small hive beetle in Italy. *Bee World*, 91(4), 88-89. http://dx.doi.org/10.1080/0005772X.2014.11417618

Mutinelli, F., Montarsi, F., Federico, G., Granato, A., Maroni Ponti, A., Grandinetti, G., Ferrè, N., Franco, S., Duquesne, V., Rivière, M.-P., Thiéry, R., Henrikx, P., Ribière-Chabert, M. & Chauzat, M.-P. (2014). Detection of *Aethina tumida* Murray (*Coleoptera: Nitidulidae.*) in Italy: outbreaks and early reaction measures. *Journal of Apicultural Research*, 53, 569-575. http://dx.doi.org/10.3896/IBRA.1.53.5.08

Mutinelli, F., Federico, G., Carlin, S., Montarsi, F. & Audisio, P. (2015) Preliminary investigation on other Nitidulidae beetles species occurring on rotten fruit in Reggio Calabria province (south-western Italy) infested with small hive beetle (*Aethina tumida*), *Journal of Apicultural Research*, 54(3), 233-235. http://dx.doi.org/10.1080/00218839.2016.1142733

Neumann, P. & Ellis, J. D. (2008). The small hive beetle (*Aethina tumida* Murray, Coleoptera: Nitidulidae): distribution, biology and control of an invasive species. *Journal of Apicultural Research*, 47(3), 181-183. http://dx.doi.org/10.3896/IBRA.1.47.3.01

Neumann, P. & Hoffmann, D. (2008). Small hive beetle diagnosis and control in naturally infested honey bee colonies using bottom board traps and CheckMite+ strips. *Journal of Pest Science*, 81, 43-48. http://dx.doi.org/doi:10.1007/s10340-007-0183-8

Neumann, P., Evans, J. D., Pettis, J. S., Pirk, C. W. W., Schäfer, M. O., Tanner, G. & Ellis, J. D. (2013). Standard methods for small hive beetle research. In *V. Dietemann, J. D. Ellis & P. Neumann (Eds) The COLOSS BEEBOOK: Volume II: Standard methods for Apis mellifera pest and pathogen research. Journal of Apicultural Research*, 52(4), http://dx.doi.org/10.3896/IBRA.1.52.4.19

Office International des Epizooties (2014). Infestation with *Aethina tumida* (Small hive beetle) Chapter 9.4. In *Terrestrial Animal Health Code*. OIE (World Organisation for Animal Health); Paris, France. http://www.oie.int/index.php?id=169&L=0&htmfile=chapitre_aethina_tumida.htm

Office International des Epizooties (2014). Small hive beetle infestation (*Aethina tumida*) (NB: Version adopted in May 2013) Chapter 2.2.5. In *Manual of Diagnostic Tests and Vaccines for Terrestrial Animals*. OIE (World Organisation for Animal Health); Paris, France. http://www.oie.int/fileadmin/Home/eng/Health_standards/tahm/2.02.05_SMALL_HIVE_BEETLE.pdf

Palmeri, V., Scirtò, G., Malacrinò, A., Laudani, F. & Campolo, O. (2015). A new pest for European honey bees: first report of *Aethina tumida* Murray (Coleoptera Nitidulidae) in Europe. *Apidologie*, 46(4), 527-529. http://dx.doi.org/10.1007/s13592-014-0343-9

Province di Reggio Calabria e di Vibo Valentia (2013). Comuni di Rosarno - Feroleto della Chiesa - Laureana di Borrello - Rizziconi - Serrata - San Pietro di Caridà - San Calogero. Piano Strutturale Associato (P.S.A.) e Regolamento Edilizio e Urbanistico (R.E.U.). *Quadro Conoscitivo Territoriale - Indagini Geologiche Relazione Descrittiva*. 64 pp. www.comune.rosarno.rc.it/dms/Comune/PSA/relazioni/SSG_Rel.pdf

Schäfer, M. O., Pettis, J. S., Ritter, W., Neumann, P. (2008). A scientific note on a quantitative diagnosis of small hive beetles, *Aethina tumida* in the field. *Apidologie*, 39, 564-565. http://dx.doi.org/doi:10.1051/apido:2008038

Schäfer, M. O., Pettis, J. S., Ritter, W. & Neumann, P. (2010). Simple small hive beetle diagnosis. *American Bee Journal*, 150, 371-372.

Valério da Silva, M. J. (2014). The first report of *Aethina tumida* in the European Union, Portugal, 2004. *Bee World*, 91(4), 90-91.

http://dx.doi.org/10.1080/0005772X.2014.11417619

Franco Mutinelli[1], Giovanni Federico[2], Fabrizio Montarsi[1], Anna Granato[1], Claudia Casarotto[1], Gianluca Grandinetti[3] Marie-Pierre Chauzat[4,5] and Andrea Maroni Ponti[6]

[1]Istituto Zooprofilattico Sperimentale delle Venezie, NRL for beekeeping, viale dell'Universita' 10, 35020 Legnaro (Padova), Italy. E-mail: fmutinelli@izsvenezie.it
[2]Istituto Zooprofilattico Sperimentale del Mezzogiorno, Sezione di Reggio Calabria, Via Nazionale 5, 89068 San Gregorio (RC), Italy.
[3]Task Force per le Attività Veterinarie, Regione Calabria, Via S. Nicola, 88100 Catanzaro (CZ) Italy.
[4]ANSES, Honey bee Disease Unit, European Reference Laboratory for honey bee health, 105 Route des Chappes – CS 20111, 06902 Sophia Antipolis, France.
[5]ANSES, Unit of Coordination and Support to Surveillance, 14 Rue Pierre et Marie Curie, 94701 Maisons-Alfort, France.
[6]Ministero della Salute, DGSAF, via G. Ribotta 5, 00144 Rome, Italy.

THREE

Surveillance for the small hive beetle (*Aethina tumida*) in Europe

Marie-Pierre Chauzat, Marion Laurent, Mike Brown, Per Kryger, Franco Mutinelli, Sophie Roelandt, Stefan Roels, Yves Van Der Stede, Marc O. Schäfer, Stéphanie Franco, Véronique Duquesne, Marie-Pierre Riviere, Magali Ribiere-Chabert and Pascal Hendrikx

Introduction

As described in Chapter Two, the small hive beetle (SHB) was reported for the first time in Reggio Calabria, south west Italy on 5th September 2014 (Mutinelli, 2014; Mutinelli *et al.*, 2014; Palmeri *et al.*, 2015). In addition SHB adults and larvae were sent to the ANSES European Union Reference Laboratory (EURL) in Sophia-Antipolis (France) where the species was also confirmed through morphological identification and through molecular diagnostics on 17 September 2014. On 18th September, confirmation of *Aethina tumida* detection in Italy was notified to the OIE (World Organisation for Animal Health).

To date (July 2015), more than 1,900 apiaries have been inspected in Calabria and more than 500 apiaries in Sicily. *A. tumida* has been confirmed in 61 apiaries located within a 20 km radius area in two provinces of Calabria region (Reggio Calabria and Vibo Valentia) with one exception in Sicily (one apiary in the province of Siracusa, directly linked to a migratory movement back from Reggio di Calabria). 3,132 honey bee colonies have been destroyed upon discovery of *A. tumida* in the apiaries (see Chapter Two for more details). Since the resuming of apiaries visits after the winter in 2015 and up to date (July 2015), 782 apiaries have been visited in the area of Calabria and 318 in Sicily. There has been no *A. tumida* outbreak observed since December 2014.

A. tumida had previously been detected In Portugal in 2004 during a control on imported queen bees from Texas, USA. All the imported materials were destroyed and stamping out was applied to the apiaries where the imported queens had been introduced. Following the implementation of these measures, no further detection of *A. tumida* occurred in Portugal and Europe (Murilhas, 2004; Valério da Silva, 2014).

To date the populations of SHB in Calabria are considered to be low. The spread of the beetle has been largely documented in other countries (USA particularly - see Chapter Five). However, to date in Italy, it is not entirely clear whether all the tracing of apiary/colony movements have been fully covered, particularly out of the surveillance area. The Italian veterinary services together with the Italian beekeepers have worked, and still work, very hard to contain SHB in Calabria by killing thousands of colonies to prevent its definite establishment and/or further dissemination in Europe. The contingency management field work in Italy is designed to reduce the risks of this new pest becoming established more widely in Europe and to maintain the SHB populations at a very low level if they might become established. These guidelines aim to provide advice to Member States on apiary surveillance and early detection of the SHB to reduce the risks of further diffusion of the beetle across other countries in Europe.

This chapter outlines the guidelines produced in 2015 by the EURL for honey bee health and the National Reference Laboratories of Belgium, Denmark, Germany, Italy and UK to support Member States with implementing a risk based framework (Chauzat *et al.*, 2015).

Spread of SHB

As documented in the USA (Hood, 2004), the spread of infestation within a territory is primarily determined by the following factors:

1. The climate and season. The biological cycle of the SHB depends on temperature and humidity conditions. Although *A. tumida* is able to withstand cold temperatures, the highest impact is usually facilitated by high temperatures and humidity (see Chapter Four).

2. The nature of the soil. Relatively moist soft, sandy soil is conducive to SHB pupation. Moisture is a limiting factor and there is less impact on colonies if not kept in the shade.

3. The density of colonies in the area. There is greater spread in areas with a high density of apiaries.

4. The structure and organisation of the beekeeping sector: areas and routes of migratory beekeeping, importing apiaries, production of package bees and/or nucleus colonies, trade in beekeeping equipment, storage of honey, etc.

European regulatory context

The SHB is a statutory notifiable pest within the European Union (European Commission, 1992). There is a legal requirement for any SHB confirmation. There is therefore a legal requirement on beekeepers to notify any suspect findings. Following introduction in Italy, protective measures have been implemented (European Commission, 2014; 2015). The dispatch of honey bees, bumble bees, unprocessed apiculture by-products, beekeeping equipment and comb honey intended for human consumption is banned from the infested regions to other areas of the Union. EU legislation prohibits (with the exception of New Zealand) imports of package bees or colonies from Third Countries. It is permitted to import honey bee queens from a very limited number of countries outside the EU (European Commission, 1992; 2010). The import regulations and protective measures are the main defence against the introduction and the spread of the SHB in Europe. It is therefore crucial that every Competent Authority, and indeed every beekeeper, respect the EU legislation and ensure regular surveillance.

Objectives of the surveillance

In a country where SHB is considered absent (still an exotic threat), the objectives of the surveillance programme could be to: detect any SHB infestation at an early stage in order to eradicate it; and demonstrate freedom from SHB infestation to maintain the country's infestation-free status. This objective should be specified in relation to the European or national regulations and international standards (OIE), particularly regarding official criteria for recognition of this status.

For an infested country, the objectives are compartmentalization and zoning to: demonstrate the absence of SHB infestation for maintaining infestation-free status in certain zones/compartments; and detect any SHB infestation at an early stage in order to eradicate it from infested zones.

Surveillance methods

Early detection can be ensured by combining extended outbreak surveillance (covering the entire national territory or surveillance zone) with passive surveillance (outbreaks) and active surveillance (programmes) targeting at-risk zones. For the entire surveillance programme, the epidemiological unit considered is the apiary, which can contain one or more colonies. A beekeeper can own more than one apiary. To identify the numbers of colonies to be inspected within each apiary according to the size of the apiary and the expected prevalence and diagnostic sensitivity, please refer to the sample size calculator (Tables 1-4).

Outbreak surveillance

Enhanced passive outbreak surveillance is based on the reporting of suspected cases by beekeepers (or any other stakeholders in the beekeeping sector) to competent veterinary authorities (European Commission, 1982; 1992). This surveillance covers all of the apiaries throughout the national territory and must therefore be promoted by the competent veterinary authority in the whole sector using all existing (in-) formal communication channels. Reporting criteria are based on the definition of a suspected case. This outbreak surveillance can be strengthened in at-risk zones meeting the criteria given below.

Active surveillance

Active surveillance involves the sampling of apiaries in which investigations are being undertaken (Fig. 1). This is suitable for infested as well as SHB free countries. This sampling can be designed in various ways, depending on the objective targeted in the Member State in terms of precision and accuracy (precision evaluates the dispersal of the measures; the accuracy refers to the systematic errors) and the means dedicated to the surveillance (particularly trained personnel and resources). The proposals given below are ranked from the most robust methodology through to lighter touch approaches.

1. Targeted sampling: selection of at-risk apiaries with a particular risk of being infested (based on beekeeping practices) for the early detection of infestation.

2. Representative sampling of all registered apiaries located in a zone considered at-risk for the early detection of infestation.

3. Representative sampling of all registered apiaries in

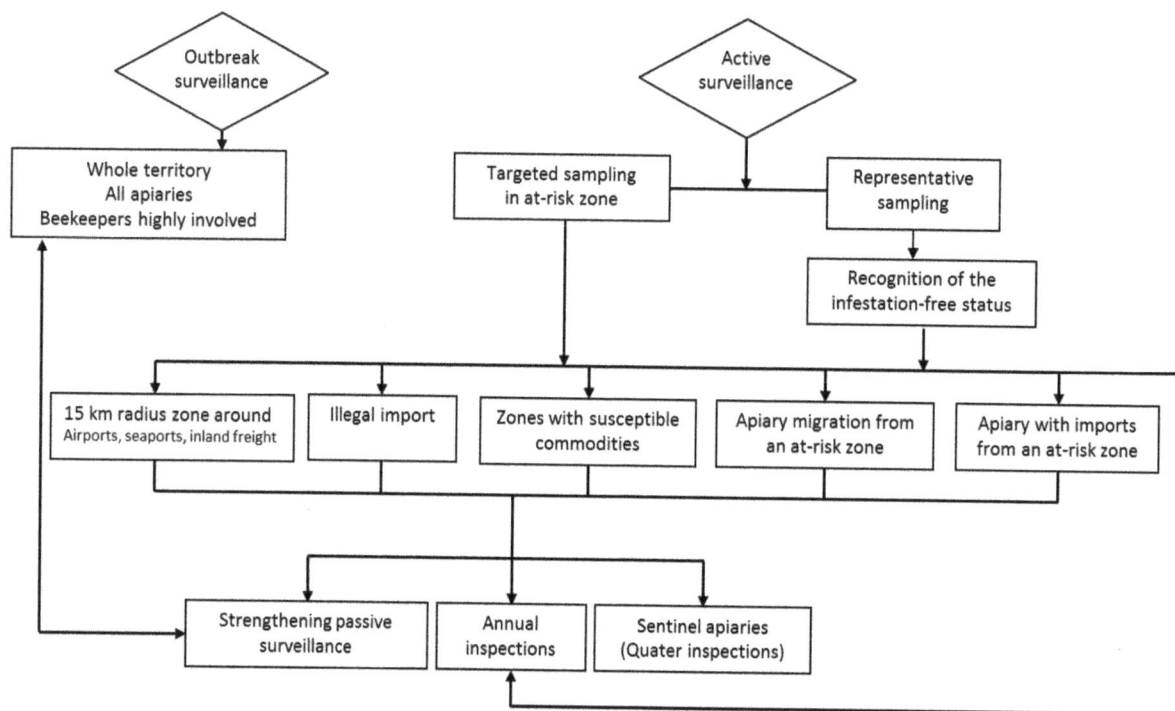

Fig. 1. Flow chart with different types of surveillance and sampling level required.

part or all of the national territory for recognition of the infestation-free zone/country status, if this is the regulatory objective.

Targeted sampling (selection) of at-risk apiaries

The individual criteria for the risk of apiary infestation are as follows: apiaries that have been moved in the last 12 months to or from a protection or surveillance zone, or a zone classified as such within 12 months of their migration; apiaries that imported, in the last 12 months, queens, swarms or package bees from a zone recognised as infested or classified as such within 12 months of importation.

These apiaries must be fully recorded and inspected as soon as possible after their identification as described below. Further inspection can be planned according to estimated level of risk and to the specific situation of the area they moved from, according to events that might shed a different light on risk factors. An event is any new epidemiological information such as outbreaks retrospectively reported in new area, tracing of movements, or field information.

Representative sampling in at-risk zones

The criteria to classify an at-risk zone are as follows: a 15-km-radius zone around an international seaport or international airport where at-risk products are imported: bee products (bees, queens, brood, hive products); beekeeping products (beekeeping equipment); other products containing bees in the broad sense (bumble bee *Bombus* sp. colonies or queens); ripe fruits or vegetables (e.g. apples and bananas); or potted plants. In addition it should be noted that, in many cases, a great deal of freight does not get opened or dealt with at ports but is moved inland into freight depots.

Each Member State will evaluate other risk hotspots as required, for example if imports occur through different pathways than seaports. Illegal imports should also be included as much as possible. But by their very nature they are illegal, and therefore unlikely to be discovered. A Member State might not even know that any has occurred and might therefore not to be able to target. However, Member States should always be on the lookout. Zones into which susceptible goods are moved e.g., by road or rail from identified surveillance or at-risk areas and protection zones. Each at-risk zone in the territory must be identified and all the apiaries located in this zone georeferenced.

Three types of actions can be taken in the apiaries in these zones in the following order of priority, considering that, depending on the situation, only one or two actions can be implemented:

1. Strengthening passive surveillance: a specific communication strategy is used to inform beekeepers, bee health workers and other beekeeping workers of the risk of SHB infestation and provide them with the knowledge required to detect suspicions of SHB infestation (European Union Reference Laboratory, 2015).

2. Quarterly inspections of three large sentinel apiaries (minimum 10 colonies) placed near the maximum risk source (maximum 15 km from seaport, airport, road, and rail network). This data are given as an indicative basis (number of apiaries and the number of colonies per sentinel apiary) and should be refined according to further information, e.g., modelling tools and epidemiological knowledge.

3. Annual inspection (preferably late spring depending on the climate of each Member State) of a random sample of registered apiaries to be able to detect infestation affecting at least 5% of the apiaries with a 95% confidence level of detection. Irrespective of the population size of the apiaries in the defined at-risk zone, the maximum sample size requested is 59 apiaries.

Representative sampling throughout the national territory (according to the regulations)

The following recommendations are given on an indicative basis and have to be refined according to upcoming national or European regulations regarding the recognition of infestation free zones or country. The national territory can be divided into agro-ecological or geographical zones with a homogeneous risk of infestation (example: beekeeping carried out in mountainous areas will be different from dry prairies or oceanic zones). Each Member state will have full knowledge of the structure of their honey bee sector for hierarchical purposes. Each of these zones should be sampled to be able to detect SHB infestation at minimum prevalence level of 2% of the apiaries with a 95% confidence level. Irrespective of the population size of the apiaries in each agro-ecological zone, the maximum sample size requested is 149 apiaries. The relevance of these sampling criteria should be determined in relation to the national or international regulations establishing prevalence limits for detection. This stratification is to ensure that, in case of large Member states, the entire country is not considered as only one population exposed to the same risk. However, if further work demonstrates that the sensitivity of the detection process is not 100%, the maximum number for a design prevalence of 2% would be greater than 149 (for example, it would be 186 for a technique with a 80% sensitivity).

Practical aspects of apiary inspection

In each selected apiary in the surveillance sample, a certain number of colonies need to be inspected in order to be able to detect infestation affecting at least 5% of the colonies with a 95% confidence level (this means that,

irrespective of the size of the apiary, no more than 59 colonies per apiary should be inspected if the sensitivity of the investigation technique reaches 100%). For small apiaries (n<20), all colonies must be inspected (Tables 1-4).

Data given in this document regarding design prevalence and sample size are based on the mathematical assumption that the investigation techniques have 100% sensitivity. It has to be acknowledged that investigation/ diagnostic techniques at colony level do not reach 100%, and most likely ranges between 90 and 95%. However, at the apiary level (the epidemiological unit), considering that in case of an infestation, more than one colony is infested, it can be assumed that the sensitivity of the investigation technique is to all intents and purposes 100%.

A colony is inspected as follows:

1. Visual observation of frames: SHB detection by observation of frames should take into account the lucifugous nature of the adults. Examination on sunny days (or with any light exposure) is recommended, since the adult beetles will scurry quickly from the light. The frames should be removed from the hive one by one. Each side of the frame should be quickly observed. The beetles tend to move rapidly along the frame to find a refuge from light and can easily be spotted by an attentive observer (Ward et al., 2007).

2. Detection of suspect specimens: if adult insects or larvae are detected during the visual inspection, their

Tables 1 to 4 (opposite). Sample size calculator indicating the number of colonies to be visited in an apiary according to the size of the apiary and to the prevalence targeted.

Table 1. Number of units to be inspected in order to detect a prevalence of 2% with a 100% sensitivity technique.

Total number of units (apiaries) in the zone	50	100	200	300	400	500	600	700	800	900	1000	1500	2000	3000	3500	4000	5000	9000	>35000
To be inspected	48	78	105	117	124	129	132	134	136	137	138	142	143	145	146	146	147	148	149

Table 2. Number of units to be inspected in order to detect a prevalence of 5% with a 100% sensitivity technique.

Total number of units (apiaries) in the zone	50	100	200	300	400	500	600	700	800	900	1000	1500	2000	>4500
To be inspected	35	45	51	54	55	56	56	57	57	57	57	58	58	59

Table 3. Number of colonies to be inspected in order to detect a prevalence of 5% with a 95% sensitivity technique.

Total number of colonies within the selected apiary	up to 24	25	30	40	50	60	70	80	100	110	120	140	160	170	200	220	300	400	500
To be inspected	all	24	28	33	37	40	42	44	47	48	49	51	52	53	54	55	56	58	59

Table 4. Number of colonies to be inspected in order to detect a prevalence of 10% with a 95% sensitivity technique.

Total number of colonies within the selected apiary	up to 13	14	15	16	18	21	23	26	29	33	38	44	52	62	77	98	134	204	>410
To be inspected	all	13	14	15	16	17	18	19	20	21	22	23	24	25	26	27	28	29	30

characteristics should be compared to the definition of a suspected case (see below) to be able to rule out obvious negative cases (e.g., wax moth larvae such as *Galleria mellonella* or *Achroia grisella*). If adult insects or larvae correspond to the definition of a case, a sample should be taken and sent to a reference laboratory for identification.

3. Collection of suspect specimens: to capture adult beetles, it is preferable to use a mouth aspirator. Once they have been captured, it is advisable to promptly kill the individuals in a container such as a sampling tube filled with alcohol (avoid the use of denatured alcohol) to keep them from flying away when the container is opened.

Traps can be placed in colonies and used in combination with the visual observation method to increase the likelihood of detection, or as an alternative to visual observation when climatic conditions do not allow colony inspection. However, in this condition traps are much less effective. Thus, depending on the zone or surveillance method, it may be decided to use visual observation or traps, but whenever possible, the combination is best. In apiaries where inspections are frequently undertaken (sentinel apiaries), surveillance may be more acceptable if traps are used. For single inspections (annual sampling), it may be best to perform visual inspections because of the high detection sensitivity and to avoid a return visit to check the trap. However, attention needs to be paid to the conditions in which traps are used (Neumann *et al.*, 2013).

Detection through molecular methods should be validated blind in the field, in natural conditions to validate the sensitivity of the method (determination of the detection limit) and to standardise the sampling frame in the purpose of detection (Ward *et al.*, 2007, Cepero *et al.*, 2014). Further experimental and field work is needed to implement these steps of validation.

Surveillance period

Surveillance by visual observation of beetles inside hives depends on the temperature conditions. When temperatures are low, the examination of hives can jeopardise the survival of the colony, which will be 'clustered'. Surveillance using floor trapping methods can be undertaken all year long without endangering bee colonies. However, it is necessary to take into account the decreased sensitivity of traps made of corrugated plastic when bee colonies are clustered (see above).

Lastly, it is important to adapt the surveillance period and methods to the expected spread of SHB. The biological cycle of this parasite depends on the temperature and humidity conditions; moreover, movements of live bees and beekeeping equipment are factors in the spread of the beetle. Surveillance should be strengthened from spring to autumn, during the active beekeeping season. It will however be necessarily less intensive in the winter, particularly in colder climates.

Definition of a case

A suspected case is defined by at least one of the following situations detected upon observation by the beekeeper or someone inspecting the apiary: 1. Occurrence in the hive (or beekeeping equipment) of one or more beetles similar to *A. tumida*; 2. Occurrence in the hive or in the hive's immediate environment (larvae climbing out of the hive to pupate in the soil / 'wandering larvae') of one or more whitish beetle-like larvae similar to *A. tumida* (different from the larvae of wax moths); 3. Occurrence of at least one beetle in a trap placed inside the hive.

Confirmation of an initial outbreak located in a zone considered non-infested: (i.e. in the framework of the surveillance plan)

A case of SHB infestation is confirmed based on at least one of the following criteria: 1. Identification of an adult SHB (*A. tumida*) by the NRL based on morphological criteria, confirmed if needed by molecular identification (e.g. damaged specimen). The EU RL for honey bee health is currently validating a molecular technique to identify *A. tumida* adults and larvae. As soon as the full procedure is ready, it will be made public and freely available to use; 2. Identification of an SHB larva by the NRL based on morphological criteria, systematically confirmed by molecular identification.

Confirmation of the following cases occurring in protection or surveillance zones established around the confirmed initial outbreak

A case of *A. tumida* infestation is confirmed based on at least one of the following criteria: 1. Identification of an adult SHB (*A. tumida*) by a the NRL based on morphological criteria, confirmed if needed by molecular identification (e.g. in the case of damaged specimens); 2. Identification of an SHB (*A. tumida*) larva by the NRL based on morphological criteria. Once an outbreak has been officially recognised, larval identification can be confirmed based only on morphological criteria in this outbreak, systematic molecular analysis being no longer necessary.

Sampling

It is important to sample as many specimens as possible (adults and larvae). Morphological identification is even more reliable if done on undamaged specimens (specimens whose morphological integrity has been preserved, have not been crushed and are in a good state of preservation). This is why the use of a mouth aspirator is recommended for sampling adult beetles, and the use of flexible entomological tweezers is recommended for sampling larvae.

All specimens must be killed before being transported. The use of 70% undenatured ethanol is recommended. Place the specimen(s) in a tube containing ethanol and tightly close it. The duly labelled container can thus be sent through the regular or specific post services, in a timely fashion to the laboratory at ambient temperature with the related background data (number of samples, etc.). It is also recommended to take pictures of suspicious signs observed in colonies, and collect specimens in appropriate sample containers and promptly send them to the NRL so that the alert level may be assessed. Pictures can be emailed directly to the Member State NRL. The laboratory should be notified by telephone and email of the shipping of samples, so it can be ready to receive and quickly analyse them.

Organisation of surveillance

The organisation of the central intermediate and field levels, the data management (collection, transmission, centralisation and validation), the communication and the training should be specifically addressed by each Member States in order to correctly organise their country surveillance systems.

References

Cepero, A., Higes, M., Martinez-Salvador, A., Meana A. & Martin-Hernandez, R. (2014). A two year national surveillance for Aethina tumida reflects its absence in Spain. BMC Research Notes, 7, 878. http://dx.doi.org/10.1186/1756-0500-7-878

Chauzat, M.-P., Laurent, M., Brown, M., Kryger, P., Mutinelli, F., Roelandt, S. & Hendrikx, P. (2015). Guidelines for the surveillance of the small hive beetle (Aethina tumida) infestation. European Union Reference Laboratory for honey bee health (EURL); Sophia-Antipolis, France. 19 pp. https://sites.anses.fr/en/minisite/abeilles/eurl-bee-health-home

European Commission. (1982). Council Directive 82/894/EEC of 21 December 1982 on the notification of animal diseases within the Community. Official Journal of the European Union, L378, 31 December 1982, pp. 58–62.

European Commission. (1992). Council Directive 92/65/EEC of 13 July 1992 laying down animal health requirements governing trade in and imports into the Community of animals, semen, ova and embryos not subject to animal health requirements laid down in specific Community rules referred to in Annex A (I) to Directive 90/425/EEC. Official Journal of the European Union, L268: 14 September 1992, pp. 54-72. [as amended by Commission Regulation (EC) No. 1398/2003 of 5 August 2003 amending Annex A to Council Directive 92/65/EEC to include the small hive beetle (Aethina tumida), the Tropilaelaps mite (Tropilaelaps spp.), Ebola and monkey pox. Official Journal of the European Union, L198: 6 August 2003, p. 3].

European Commission. (2010). Commission Regulation (EU) No. 206/2010 of 12 March 2010 laying down lists of third countries, territories or parts thereof authorised for the introduction into the European Union of certain animals and fresh meat and the veterinary certification requirements. Official Journal of the European Union, L73, 20 March 2010, 120 pp.

European Commission. (2014). Commission Implementing Decision of 12 December 2014 concerning certain protective measures with regard to confirmed occurrences of the small hive beetle in Italy. (2014/909/EU). Official Journal of the European Union, L359, 16 December 2014, pp. 161-163.

European Commission. (2015). Commission Implementing Decision of 28 May 2015 amending Implementing Decision 2014/909/EU by extending the period of application of the protection measures in relation to the small hive beetle in Italy. (2015/838/EU). *Official Journal of the European Union*, L132, 29 May 2015, pp. 86-87.

European Union Reference Laboratory. (2015) *Small hive beetle*. ANSES; France. 2 pp. https://sites.anses.fr/en/system/files/SHB_For_beekeepers_2015feb_0.pdf

Hood, M. W. M. (2004). The small hive beetle, *Aethina tumida*: a review. *Bee World*, 85(3), 51-59. http://dx.doi.org/10.1080/0005772X.2004.11099624

Murilhas, A. M. (2004). *Aethina tumida* arrives in Portugal. Will it be eradicated? *EurBee Newsletter*, 2, 7-9.

Mutinelli, F. (2014). The 2014 outbreak of the small hive beetle in Italy. *Bee World*, 91(4), 88-89. http://dx.doi.org/10.1080/0005772X.2014.11417618

Mutinelli, F., Montarsi, F., Federico, G., Granato, A., Maroni Ponti, A., Grandinetti, G., Ferrè, N., Franco, S., Duquesne, V., Rivière, M.-P., Thiéry, R., Henrikx, P., Ribière-Chabert, M. & Chauzat, M.-P. (2014). Detection of *Aethina tumida* Murray (*Coleoptera: Nitidulidae*) in Italy: outbreaks and early reaction measures. *Journal of Apicultural Research*, 53, 569-575. http://dx.doi.org/10.3896/IBRA.1.53.5.08

Neumann, P., Evans, J. D., Pettis, J. S., Pirk, C. W. W., Schäfer, M. O., Tanner, G. & Ellis, J. D. (2013). Standard methods for small hive beetle research. In *V. Dietemann, J. D. Ellis & P. Neumann (Eds) The COLOSS BEEBOOK: Volume II: Standard methods for* Apis mellifera *pest and pathogen research. Journal of Apicultural Research*, 52(4), http://dx.doi.org/10.3896/IBRA.1.52.4.19

Palmeri, V., Scirtò, G., Malacrinò, A., Laudani, F. & Campolo, O. (2015). A new pest for European honey bees: first report of *Aethina tumida* Murray (Coleoptera Nitidulidae) in Europe. *Apidologie*, 46(4), 527-529. http://dx.doi.org/10.1007/s13592-014-0343-9

Schäfer, M. O., Pettis, J. S., Ritter, W. & Neumann, P. (2008). A scientific note on a quantitative diagnosis of small hive beetles, *Aethina tumida* in the field. *Apidologie*, 39, 564-565. http://dx.doi.org/10.1051/apido:2008038

Spiewok, S. & Neumann P. (2006). Infestation of commercial bumblebee (*Bombus impatiens*) field colonies by small hive beetles (*Aethina tumida*). *Ecological Entomology*, 31, 623-628. http://dx.doi.org/10.1111/j.1365-2311.2006.00827.x

Valério da Silva, M. J. (2014). The first report of *Aethina tumida* in the European Union, Portugal, 2004. *Bee World*, 91(4), 90-91. http://dx.doi.org/10.1080/0005772X.2014.11417619

Ward, L., Brown, M., Neumann, P., Wilkins, S., Pettis, J. S. & Boonham, N. (2007). A DNA method for screening hive debris for the presence of small hive beetle (*Aethina tumida*). *Apidologie*, 38, 272-280. http://dx.doi.org/10.1051/apido:2007004

Marie-Pierre Chauzat[1,2], Marion Laurent[2], Mike Brown[3], Per Kryger[4], Franco Mutinelli[5], Sophie Roelandt[6], Stefan Roels[6], Yves Van Der Stede[6], Marc Schäfer[7], Stéphanie Franco[2], Véronique Duquesne[2], Marie-Pierre Riviere[2], Magali Ribiere-Chabert[2] and Pascal Hendrikx[1]

[1]ANSES, Unit of Coordination and Support to Surveillance, Maisons-Alfort, France.
Email: Marie-pierre.CHAUZAT@anses.fr
[2]ANSES, Honey bee Disease Unit, European Reference Laboratory for honey bee health, Sophia Antipolis, France.
[3]APHA, National Bee Unit, Sand Hutton York, YO10 4BG, UK.
[4]Århus University, Department of Agroecology, 4200 Slagelse, Denmark.
[5]Istituto Zooprofilattico Sperimentale delle Venezie, NRL for beekeeping, Legnaro (Padova), Italy.
[6]Groeselenberg 99, 1180 Brussels, Belgium.
[7]Friedrich-Loeffler-Institut, Südufer 10 17493 Greifswald-Insel Riems, Germany.

FOUR

The small hive beetle in Italy: what can we expect in the future?

Peter Neumann

Introduction

Small hive beetles (SHB) *Aethina tumida* are parasites and scavengers of honey bee, *Apis mellifera*, colonies and are native to sub-Saharan Africa, where they are usually considered to be a minor pest only (Lundie, 1940; Hepburn & Radloff, 1998; Neumann & Elzen, 2004). In the past they have been introduced mainly via import of bees and/or bee products into several new areas, e.g. USA (1996), Egypt (2000), Australia (2001) and Europe (2004; see Neumann and Ellis, 2008 for an overview). In the USA and Australia, SHB are now well established as an invasive species, and can be considered an economically significant pest of European-derived honey bees under suitable environmental conditions (Neumann & Elzen, 2004; Spiewok *et al.*, 2007). Their recent introduction into Italy (Mutinelli, 2014; Mutinelli *et al.*, 2014; Palmeri *et al.*, 2015) (see Chapter Two) has therefore raised considerable concern that SHB will also be able to establish populations in Europe.

Small hive beetle reproduction

A key question for the Italian case and any future introductions of SHB is clearly whether eradication measures such as burning of infested hives will be successful in eliminating the invasive populations. What are the chances for a successful eradication? To answer this question, the different reproductive options of small hive beetles should be considered:

1. Honey bees. In honey bee colonies, SHB can mass reproduce, often resulting in the full structural collapse of the entire nest (Hepburn & Radloff, 1998). Even reasonably strong queenright colonies of European origin can collapse in less than five days (3-5 frames of brood, 6-8 frames of bees; Neumann *et al.*, 2001). In my eighteen years of SHB field work, I have never observed that such beetle-induced collapses occur in strong honey bee colonies in Africa. This is probably due to quantitative behavioural differences between African and European honey bee subspecies, (Neumann & Elzen, 2004). SHB can also show cryptic low-level reproduction in hive debris (Spiewok & Neumann, 2006b; Fig. 1.) or underneath capped cells (Neumann & Hoffmann, 2008). Such cryptic reproduction may remain unnoticed by both bees and beekeepers because very few larvae are produced in infested colonies (Spiewok & Neumann, 2006b) without any obvious signs of beetle associated damage (personal observations). This way of reproduction is likely to maintain rather low beetle populations. Nevertheless, jettisoning workers (Lundie, 1940), which carry beetle larvae out of the hive at some

Fig. 1. Cryptic low level reproduction of SHB in the hive debris of a honey bee colony. Three roaming beetle larvae are indicated with red circles.

distance, can occasionally be seen at the entrance of such infested colonies (Fig. 2.). Moreover, apiculture can massively support SHB reproduction, e.g. via poor sanitary standards in honey houses etc. (Spiewok et al., 2007; Fig. 3.). Another option for SHB reproduction in association with honey bees are wild or feral colonies emerging from escaped swarms (Gillespie et al., 2003; Neumann et al., 2010).

2. Bumble bees (*Bombus* spp). It has long been known that SHB can also complete an entire life cycle in association with bumble bee colonies (Ambrose et al., 2000). Research since then has demonstrated that odours

from bumble bees and their hive products are attractive for adult small hive beetles (Spiewok & Neumann, 2006a; Graham et al., 2011) and that natural infestations of commercial bumble bee colonies can occur both in the field (Spiewok & Neumann, 2006a) and in glasshouses (Hoffmann et al., 2008).

3. Stingless bees. SHB have also been reported to naturally infest colonies of stingless bees in Africa (Mutsaers, 2006), Australia (cf. Greco et al., 2010; cf. Halcroft et al., 2011) and America (Peña et al., 2014).

4. Other food stuffs. Finally, successful reproduction of SHB on fruits and other food stuff (e.g. decaying beef schnitzel) has been reported in the laboratory (Ellis et al., 2002; Buchholz et al., 2008; Fig. 4.) and at low levels in the field (fruits only, Buchholz et al., 2008).

Fig. 2. Honey bee workers removing an SHB larva from the hive entrance.

In contrast to many other honey bee parasites, SHB appear to be fairly opportunistic, and probably oviposit and feed on whatever food is available. I suggest that this should be considered when evaluating the success chances of eradication programmes and pest management strategies in established beetle populations.

Eradication

Experiences from prior SHB invasions should also be considered. In the USA, SHB were only confirmed in 1998, two years after their introduction in 1996, when considerable damage to apiculture occurred in Florida (Sanford, 1998; Neumann & Elzen, 2004). This was far too late for any eradication measures to be effective (see Chapter Five). In July 2002, SHB associated damage was

Fig. 4. Mass reproduction of SHB on a banana diet. This is a cheap and feasible alternative to feeding the beetles with precious honey bee brood frames or pollen / honey paste (Neumann et al., 2013).

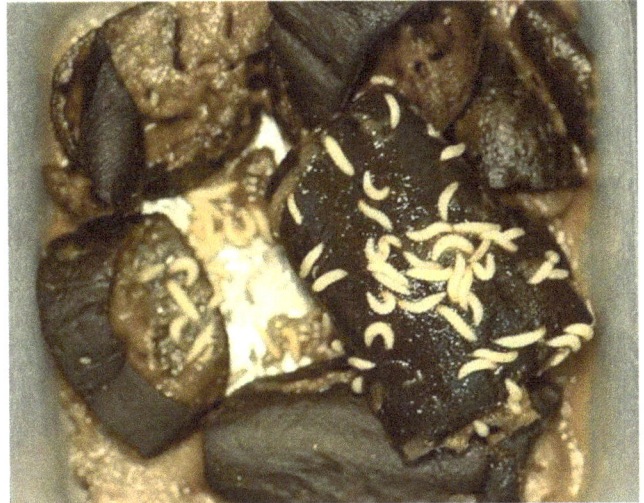

Fig. 3. Mass reproduction of SHB in a US honey house. The post-feeding wandering larvae are positive phototactic (red circles) and leave the honey house for pupation in the nearby soil.

noticed by Michael Duncan in his apiary in Richmond, New South Wales, Australia. The beetles were identified as *A. tumida* in October 2002. Since SHB reproduction was already found in feral honey bee colonies (Gillespie et al., 2003), the Australian government decided against an eradication programme (see Chapter Seven). In case of the Portugal introduction, small hive beetle eradication was successful, possibly because very few specimens were introduced and the detection was early (Murilhas, 2004; Valério da Silva, 2014). In conclusion, experience from past introductions strongly suggests that a timely response is a key factor for success of SHB eradication programmes.

In Italy, SHB pupae were also found besides the other life stages (eggs, larvae and adults; Mutinelli *et al.*, 2014). So far, 61 infested apiaries were confirmed in Calabria and Sicily, and an infested feral colony was found in Gioia Tauro municipality (Mutinelli *et al.*, 2014; Chapter Two). Taken together, these reports indicate that SHB have been present and reproducing in Italy long before their first detection in September 2014. The situation looks promising at present because no new cases have been reported between December 2014 and July 2015 (Istituto Zooprofilattico Sperimentale delle Venezie, 2015). Chances of SHB survival in Italy are, however, not zero despite the comprehensive efforts of the local stakeholders. Indeed, irrespective of the success of any eradication measure associated with apiculture, SHB are known to be able to survive outside of apiculture (see above, e.g. in feral honey bee colonies).

If SHB cannot be eradicated in Italy now or in future cases of introductions into Europe, what will be the potential consequences? Since adult SHB overwinter in the honey bee cluster (Schäfer *et al.*, 2010), they can survive in association with colonies throughout Europe, especially because they seem to be long-lived (Schmolke, 1974). A weak point in their life cycle is, however, their pupation in the soil outside of colonies (Neumann & Elzen, 2004). Indeed, the length and success of the life cycle strongly depends on soil temperature, texture and humidity (Lundie, 1940; Schmolke, 1974; Neumann *et al.*, 2001; Ellis *et al.*, 2004; Haque & Levot, 2005; Müerrle & Neumann, 2004; de Guzman & Frake, 2007). Therefore, a Mediterranean climate combined with adequate sandy soil and humidity is likely to enhance SHB reproduction compared to other European regions. Indeed, the Mediterranean climate is very similar to Southern Africa, but European honey bees appear to be more susceptible to SHB infestations (Neumann & Elzen, 2004). Since damage to honey bee colonies has been reported from regions with significantly higher SHB infestation levels (Spiewok *et al.*, 2007), the number of possible beetle generations per year appears to be a reasonable good token of potential damage.

Therefore, there will be a potential for low to high relative SHB risk in Europe based on climatic data only. While in a Nordic or temperate climate, the potential for beetle damage will be low or moderate (as in the USA and Canada), the potential risk will be higher in Southern Europe. In conclusion, as in Australia and the USA, SHB will require adequate attention by European beekeepers to limit damage such as diagnosis and control as well as adjusted management (including sanitation in honey houses; see Hood 2011 for an overview).

Acknowledgements

Appreciation is addressed to Marco Lodesani and Cecilia Costa for excellent local organization of the small hive beetle COLOSS workshop held in Bologna in 2015, and to Gina Retschnig for constructive comments on an earlier draft of the manuscript. Financial support was granted by the Ricola Foundation Nature and Culture and the Vinetum Foundation.

References

Ambrose, J. T., Stanghellini, M. S. & Hopkins, D. I. (2000). A scientific note on the threat of small hive beetles (*Aethina tumida* Murray) to bumble bee (*Bombus* sp.) colonies in the United States. *Apidologie*, 31, 455-456.

Buchholz, S. B., Schäfer, M. O., Spiewok, S., Pettis, J. S., Duncan, M., Ritter, W., Spooner-Hart, R. & Neumann, P. (2008). Alternative food sources of *Aethina tumida* (Coleoptera: Nitidulidae). *Journal of Apicultural Research*, 47(3), 202-209.
http://dx.doi.org/10.3896/IBRA.1.47.3.08

de Guzman, L. I. & Frake, A. M. (2007). Temperature affects *Aethina tumida* (Coleoptera: Nitidulidae) Development. *Journal of Apicultural Research*, 46(2), 88-93.
http://dx.doi.org/10.1080/00218839.2007.11101373

Ellis, J. D., Neumann, P., Hepburn, H. R. & Elzen, P. J. (2002). Longevity and reproductive success of *Aethina tumida* (Coleoptera: Nitidulidae) fed different natural diets. *Journal of Economic Entomology*, 95, 902-907.

Ellis, J. D., Hepburn, R., Luckman, B. & Elzen, P. J. (2004). Effects of soil type, moisture, and density on pupation success of *Aethina tumida* (Coleoptera: Nitidulidae). *Environmental Entomology*, 33(4), 794-798.

Gillespie, P., Staples, J., King, C., Fletcher, M. J. & Dominiak, B. C. (2003). Small hive beetle, *Aethina tumida* (Murray) (Coleoptera: Nitidulidae) in New South Wales. *General and Applied Entomology*, 32, 5-7.

Graham, J. R., Ellis, J. D., Carroll, M. J. & Teal, P. E. A. (2011). *Aethina tumida* (Coleoptera: Nitidulidae) attraction to volatiles produced by *Apis mellifera* (Hymenoptera: Apidae) and *Bombus impatiens* (Hymenoptera: Apidae) colonies. *Apidologie*, 42, 326-336.

Greco, M. K., Hoffmann, D., Dollin, A., Duncan, M., Spooner-Hart, R. & Neumann, P. (2010). The alternative Pharaoh approach: stingless bees mummify beetle parasites alive. *Naturwissenschaften*, 97, 319-323.

Halcroft, M., Spooner-Hart, R. & Neumann, P. (2011). Behavioural defence strategies of the stingless bee, *Austroplebeia australis*, against the small hive beetle, *Aethina tumida. Insectes Sociaux*, 58, 245-253.

Haque, N. M. M. & Levot, G. W. (2005). An improved method of laboratory rearing the small hive beetle *Aethina tumida* Murray (Coleoptera: Nitidulidae). *Entomologia Generalis / Journal of General Applied Entomology*, 34, 29-33.

Hepburn, H. R. & Radloff, S. E. (1998). *Honey bees of Africa*. Springer Verlag; Berlin, Germany.

Hoffmann, D., Pettis, J. S. & Neumann, P. (2008) Potential host shift of the small hive beetle (*Aethina tumida*) to bumble bee colonies (*Bombus impatiens*). *Insectes Sociaux*, 55, 153-162.

Hood, W. M. (2011). Handbook of small hive beetle IPM. Clemson University Cooperative Extension Program. *Extension Bulletin*, 160. 20 pp.
http://www.extension.org/sites/default/files/Handbook_of_Small_Hive_Beetle_IPM.pdf

Istituto Zooprofilattico Sperimentale delle Venezie (2015) *Aethina tumida* in Italy: updates. http://www.izsvenezie.com/aethina-tumida-in-italy/

Lundie, A. E. (1940). The small hive beetle, *Aethina tumida*. *Science Bulletin Union of South Africa*, 220, 5-19

Murilhas, A. M. (2004) *Aethina tumida* arrives in Portugal. Will it be eradicated? *EurBee Newsletter*, 2, 7-9.

Műerrle, T. M. & Neumann, P. (2004). Mass production of small hive beetles (*Aethina tumida* Murray, Coleoptera: Nitidulidae). *Journal of Apicultural Research*, 43(3), 144-145.

http://dx.doi.org/10.1080/00218839.2004.11101125

Mutinelli, F. (2014). The 2014 outbreak of the small hive beetle in Italy. *Bee World*, 91(4), 88-89.

http://dx.doi.org/10.1080/0005772X.2014.11417618

Mutinelli, F., Montarsi, F., Federico, G., Granato, A., Maroni Ponti, A., Grandinetti, G., Ferrè, N., Franco, S., Duquesne, V., Rivière, M.-P., Thiéry, R., Henrikx, P., Ribière-Chabert, M. & Chauzat, M.-P. (2014). Detection of *Aethina tumida* Murray (*Coleoptera: Nitidulidae.*) in Italy: outbreaks and early reaction measures. *Journal of Apicultural Research*, 53, 569-575.

http://dx.doi.org/10.3896/IBRA.1.53.5.08

Mutsaers, M. (2006). Beekeepers observations on the small hive beetle (*Aethina tumida*) and other pests in bee colonies in West and East Africa. In *Proceedings of the 2nd European Conference of Apidology, Prague, Czech Republic.* p 44.

Neumann, P. & Ellis, J. D. (2008). The small hive beetle (*Aethina tumida* Murray, Coleoptera: Nitidulidae): distribution, biology and control of an invasive species. *Journal of Apicultural Research*, 47(3), 181-183.

http://dx.doi.org/10.3896/IBRA.1.47.3.01

Neumann, P., & Elzen, P. J. (2004). The biology of the small hive beetle (*Aethina tumida*, Coleoptera : Nitidulidae): gaps in our knowledge of an invasive species. *Apidologie*, 35(3), 229-247.

Neumann, P., Evans, J. D., Pettis, J. S., Pirk, C. W. W., Schäfer, M. O., Tanner, G. & Ellis, J. D. (2013). Standard methods for small hive beetle research. In *V. Dietemann, J. D. Ellis & P. Neumann (Eds) The COLOSS BEEBOOK: Volume II: Standard methods for Apis mellifera pest and pathogen research. Journal of Apicultural Research*, 52(4), http://dx.doi.org/10.3896/IBRA.1.52.4.19

Neumann, P. & Hoffmann, D. (2008). Small hive beetle diagnosis and control in naturally infested honey bee colonies using bottom board traps and CheckMite+ strips. *Journal of Pest Science*, 81, 43-48.

Neumann, P., Hoffmann, D., Duncan, M., Spooner-Hart, R. & Pettis, J. S. (2012). Long-range dispersal of small hive beetles. *Journal of Apicultural Research*, 51(2), 214-215.

http://dx.doi.org/10.3896/IBRA.1.51.2.11

Neumann, P., Pirk C. W. W., Hepburn, H. R., Elzen, P. J. & Baxter, J. R. (2001). Laboratory rearing of small hive beetles *Aethina tumida* (Coleoptera, Nitidulidae). *Journal of Apicultural Research*, 40(3-4), 111-112.

http://dx.doi.org/10.1080/00218839.2001.11101059

Palmeri, V., Scirtò, G., Malacrinò, A., Laudani, F. & Campolo, O. (2015). A new pest for European honey bees: first report of *Aethina tumida* Murray (Coleoptera Nitidulidae) in Europe. *Apidologie*, 46(4), 527-529. http://dx.doi.org/10.1007/s13592-014-0343-9

Peña W.L., Carballo L.F., Lorenzo J.D. (2014). Reporte de *Aethina tumida* Murray (Coleoptera, Nitidulidae) en colonias de la abeja sin aguijón *Melipona beecheii* Bennett de Matanzas y Mayabeque. *Revista de Salud Animal*, 36(3), 201-204. ISSN 0253-570X

Sanford, M. T. (1998). *Aethina tumida*: a new bee hive pest in the Western Hemisphere. *APIS (University of Florida)*, 16 (7), 1-5.

Schäfer, M. O., Ritter, W., Pettis, J. S. & Neumann, P. (2010) Winter losses of honey bee colonies (Hymenoptera: Apidae): the role of infestations with *Aethina tumida* (Coleoptera: Nitidulidae) and *Varroa destructor* (Parasitiformes: Varroidae). *Journal of Economic Entomology*, 103, 10-15.

Schmolke, M. D. (1974). A study of *Aethina tumida*: the small hive beetle, *Project Report, University of Rhodesia*. pp. 178.

Spiewok, S. & Neumann, P. (2006a). Infestation of commercial bumble bee (*Bombus impatiens*) field colonies by small hive beetles (*Aethina tumida*). *Ecological Entomology*, 31, 623-628. http://dx.doi.org/10.1111/j.1365-2311.2006.00827.x

Spiewok, S. & Neumann, P. (2006b). Cryptic low-level reproduction of small hive beetles in honey bee colonies. *Journal of Apicultural Research*, 45(1), 47-48. http://dx.doi.org/10.1080/00218839.2006.11101313

Spiewok, S., Pettis, J. S., Duncan, M., Spooner-Hart, R., Westervelt, D. & Neumann, P. (2007). Small hive beetle, *Aethina tumida*, populations I: Infestation levels of honey bee colonies, apiaries and regions. *Apidologie*, 38, 595-605.

Valério da Silva, M. J. (2014). The first report of *Aethina tumida* in the European Union, Portugal, 2004. *Bee World*, 91(4), 90-91. http://dx.doi.org/10.1080/0005772X.2014.11417619

Villa, J. D. (2004). Swarming behavior of honey bees (Hymenoptera: Apidae) in south eastern Louisiana. *Annals of the Entomological Society of America*, 97, 111-116.

Peter Neumann[1,2]

[1]Institute of Bee Health, Vetsuisse Faculty, University of Bern, Bremgartenstr. 109a, CH-3001 Bern, Switzerland. Email: peter.neumann@vetsuisse.unibe.ch
[2]Social Insect research Group, Department of Zoology & Entomology, University of Pretoria, Private Bag X20, Hatfield 0028, Pretoria, South Africa.

FIVE

Small hive beetles in the Americas

Jeff S Pettis, Marc O Schäfer and Peter Neumann

Introduction

The first detection of small hives beetles (SHB) outside of their native range in Africa was in the USA in 1996. Since this introduction into the USA the beetle has continued to spread to other areas of the Americas including Canada and parts of Central America (See Chapter Four). The impact of this pest has been most severe in warm moist climates, and has caused beekeepers to change many management practices. Management changes include; rapid extraction of harvested honey, making new colony divisions or nuclei with a high number of adult bees relative to brood, and when rearing queens, the small mating nucleus colonies are particularly vulnerable to SHB damage and require constant vigilance and or treatment for adult SHB. As SHB continue their march across the Americas, beekeepers must adapt, to avoid costly damage to bee colonies and honey. In this chapter we will provide a timeline of the movement of the beetle in the Americas and point out means that beekeepers use to deal with this new pest. Please refer to Hood (2004), Neumann and Elzen (2004), Ellis and Munn (2005), Ellis and Hepburn (2006) and Chapter Four as sources of references for more detailed biology and control of SHB.

How do beetles let you know they are present?

The first thing that most beekeepers notice when SHB enters a new area or country is the damage they cause to honey awaiting extraction (Fig. 1b). If honey is not extracted in two to three days after removal from the hive, beetles will feed on the pollen and or brood in the honey frames and make the honey unfit for extraction due to the "slime" that the beetle larvae leave as they move across the frames (Fig. 1a). Honey combs containing stored pollen and or brood can be utilized by SHB to reproduce, and once removed from the bees these honey combs are soon overrun with larvae. Additionally, even if the honey is extracted, the supers containing the remaining honey ("wet supers") can allow SHB to reproduce if small amounts of pollen and or brood are present (Hood, 2011). Lastly, any combs, wax or debris that are present in the beekeeping facility could serve as a source for SHB reproduction. General sanitation in beekeeping facilities was a lesson that beekeepers learned very rapidly once SHB were found in the USA. Rapid honey extraction and honey house sanitation are a major tool to limit the buildup of SHB populations in any given area.

After the damage noticed in honey awaiting extraction, the second most likely place that beekeepers will notice SHB is in the apiary, where adult beetles can be seen inside the colonies (Fig. 2a) or beekeepers will find colonies or nuclei that contain SHB larvae. When SHB mass reproduction occurs in a colony (Neumann and Elzen,

2004), the bees will abandon the hive and the larvae can be seen in large masses (Fig. 1a) in the combs or on the bottom boards. When SHB first arrives in an area the low levels of beetles allow them to often go undetected for several years, as was the case in the USA (cryptic low-level reproduction; Spiewok and Neumann, 2006). It was a minimum of two years (1996-1998), and probably longer, that SHB went undetected in the south eastern USA following introductions by, as yet unidentified means, into South Carolina and Florida. A variety of traps and detection techniques are now available to monitor for SHB presence (see Chapter Four), but SHB are most likely to first announce their presence by sliming stored honey combs or damaging colonies in the apiary.

A timeline of SHB in the USA

The SHB went undetected officially for at least two years. The first unidentified beetles were collected in November 1996, in Charleston, South Carolina (Hood, 2000) and later identified as *Aethina tumida*. In only two years, SHB were well established in Florida and caused considerable damage to apiculture (~$3 million in 1998 only; Ellis *et al.*, 2002c). The beetles were confirmed to be *A. tumida* in St. Lucie, Florida (June 1998; Hood, 2000) and soon found to be widespread, so it was far too late for eradication and quarantine methods to be put in place. Since then, SHB has become well established across the continental USA with all 48 contiguous states having had some positive finds or colonies with beetles transported within the state (J. Pettis,

pers. obs.). However, higher SHB populations are always found in states with warm moist climates such as Florida and Georgia. Damage by SHB can be severe if environmental conditions and/or poor management allow beetle populations to build. A survey on managed colony losses in the winter of 2006–2007 revealed that commercial beekeepers believed invertebrate pests (*Varroa destructor, Acarapis woodi* and/or SHB) were the leading cause of colony mortality (Van Engelsdorp *et al.*, 2007). As a general rule, beekeepers in the US have learned to limit the impact and build-up of SHB with improved management techniques. The major area of

Fig. 1. SHB can be a major problem in the honey house, where larvae (a) can feed on pollen or brood in the honey combs because adult bees have been removed as supers are removed from the hive and brought indoors (b) for extraction. (Photos: Jeff S Pettis USDA-ARS).

beekeeping that has remained problematic is in queen rearing, where small mating nucleus colonies (Fig. 2c) are very vulnerable to SHB damage (Mustafa *et al.,* 2014).

Observations on seasonality and dispersal in the USA

SHB reproduction is limited to times of the year when soil temperatures allow the beetle to complete its life cycle (above 16^{0}C). Beetle population build-ups have varied from Florida where they were highest in the spring and summer months, peaking in May and June (Arbogast *et al.,* 2010) and in Louisiana, peaks were noted in September and

Fig. 2. SHB adults (a) are about 1/3 the size of worker bees and can be easily transported to new areas with migratory beekeeping (b). Small queen mating nuclei (c) are particularly at risk of small hive beetle damage, due in part to their small size and low number of adult bees to protect the combs (Photos: Jeff S Pettis USDA-ARS).

November (de Guzman *et al.,* 2010). Similarly, in Georgia, SHB populations rose in July-August and peaked in September-October (Berry, 2009). In Louisiana, SHB abundance was correlated with hot days, but not with cool, dry, or humid days, or with rainfall (de Guzman *et al.,* 2010).

Adult SHBs can easily fly and are thought to be capable of flying long distances (>10km; Neumann and Elzen, 2004). However, the real distance that SHB can fly is currently unknown. Apiary density, SHB population and SHB mass reproduction determine SHB infestation levels in newly installed apiaries (Spiewok *et al.,* 2007, 2008). Male SHB were shown to fly earlier than females in the U.S. (Neumann and Elzen, 2004), but this was not found in Australia and Africa (Spiewok and Neumann, 2012).

How did SHB disperse in North America?

The rapid movement of SHB in the USA is likely to have been the result of migratory beekeeping (Fig. 2a), and the movement of package bees (Fig. 3) and beekeeping equipment (Neumann and Elzen, 2004; Gordon *et al.,* 2014; Annand, 2011). In the USA, the more southern States (North Carolina, Georgia, South Carolina and Florida) became infested between 1996 and 1998, and then the states bordering Canada only two years later in 2000 (Neumann and Elzen, 2004). This large jump in dispersal is best explained by migratory beekeeping, since many managed colonies overwinter in Florida before being transported north for pollination in the Spring (Pettis *et al.,*

2014; see also Australia: Graham *et al.*, 2014; Chapter Seven).

SHB made a huge leap to Hawaii in April 2010, where a beekeeper on the Big Island found adult beetles in hives that were confirmed to be SHB (Robson, 2012). Since then, SHB has become well established and spread from the Big island to Oahu (2010), Molokai (2011) and Kauai (2012; Martin, 2013). Beetles have had dramatic negative effects on the local honey bee industry with European-derived colonies (Connor, 2011a). In a survey of managed colonies, 55 % were reported dead with 80 % of losses attributed by the beekeepers to SHB or a combination of *V. destructor* and SHB (29 %; Connor, 2011a). Numerous feral honey bee

Fig. 3. Package bees being installed in Florida, USA. Package bees and migratory beekeeping are two major routes for the spread of small hive beetles to new areas (Photo: Denis vanEngelsdorp).

colonies (thanks to the absence of *V. destructor* before 2010) may have served as a reproductive source for large numbers of beetles (Connor, 2011a) similar to Australia (Neumann *et al.*, 2012; Chapter Seven). Finally, the environmental conditions on Hawaii are perfectly suited for SHB pupation (Connor, 2011b). The establishment of SHB has resulted in lost export markets to the queen bee industry due to quarantine restrictions in some countries (Robson, 2012). Besides Hawaii, the areas most severely affected by the SHB have been Florida and the south eastern USA (Neumann and Elzen, 2004) and queen breeders in various states (J. Pettis, *pers. obs.*). So far, there are no confirmed reports of SHB in Alaska or Puerto Rico. Canada has had limited SHB outbreaks; 2002 in Manitoba, 2006 in Alberta and Manitoba, 2008 and 2009 in Quebec, and Ontario in 2008 and 2013) (Clay, 2006; Neumann and Ellis, 2008; Giovenazzo and Boucher, 2010; Kozak, 2010; Dubuc, 2013). Ontario has had an established beetle population in Essex County since 2010. When SHB is found elsewhere in the province, the infested colonies are either killed or transported inside the quarantine area of Essex (Dubuc, 2013). Movement of bees from Ontario to Quebec or all other Canadian provinces is under strict surveillance. Ongoing surveys along the Quebec-USA border over the past six years have not detected SHB since 2012 (Giovenazzo and Bernier, 2015), with the exception of a single case in Quebec, in 2013 (Dubuc, 2013). British Columbia, the most western Canadian province with an almost Mediterranean climate appears to be SHB-free. SHB is not well established in Canada (except Ontario), probably thanks to unfavourable climatic conditions, but the commercial impact (trade regulations, movement restrictions, etc.) can be problematic for local beekeepers.

Central America and the Caribbean

Mexico first reported in 2007 (Del Valle Molina, 2007) and the beetle is now well established in at least eight Mexican states. In the tropical states (e.g. Yucatan), infestation levels can be extremely high, with hundreds of beetles per hive reported (Loza et al., 2014). This is surprising since the local honey bees are Africanized and thought to be less susceptible to SHB infestations. The underlying reasons remain unclear, and more data from Mexico are needed to understand the factors leading to high SHB infestations levels. What is happening in Mexico is also relevant for the tropical parts of central and South America.

SHB were detected in El Salvador in 2013 (Arias, 2014). A follow-up survey in 2014 detected beetles in 68 of 1,700 hives suggesting a localized outbreak (V. Landaverde, pers. comm.). Nicaragua first reported SHB in February 2014 in an area bordering Costa Rica (Gutierrez, 2014; Caldéron et al., 2015). However, it is currently unknown whether SHB is well established in Nicaragua. In Costa Rica, two SHB surveys in 2012 and 2014 were negative for SHB presence, and since then no SHB have been reported (Ramírez et al., 2014; R. A. Caldéron, pers. comm.). Since SHB are well established in the Mexican Yucatan bordering Belize and

Guatemala (Loza et al., 2014), it may be only a short time until the SHB will reach Belize and Guatemala.

SHB was found in Jamaica in 2005 (FERA, 2010) and has since spread across the island (H. Smith, pers. comm.). Although early reports suggested that the beetle can be a serious pest of local European honey bees (FERA, 2010), the well-established SHB populations do not appear to be causing problems (H. Smith, pers. comm.). This is surprising since local beekeepers appear to not use control measures other than putting hives on concrete (H. Smith, pers. comm.). The local bees are probably Africanized, but retain some European traits (e.g. gentleness, Rivera-Marchand et al., 2012); thus it is not surprising that they can deal with SHB infestations.

In Cuba, SHB was confirmed in 2012 (Milián, 2012; Darias, 2014). A. tumida is currently present in the seven provinces and is expected to extend its range island wide (Darias, 2014). So far, no major effects on local honey bees have been reported (Borroto et al., 2014), which might be due to low initial infestation rates (Spiewok et al., 2007). There are no further published reports of SHB from the Caribbean but climatic conditions are ideal and beetle spread is likely to occur.

How did the beetle spread?

Migratory beekeeping and or active dispersal of SHB appears to be crucial (Québec, Canada (Evans et al., 2003, 2008; Giovenazzo and Boucher, 2010); Coahuila, Mexico (Del Valle Molina, 2007)). Moreover, import and export of bees and bee products seem to play an important role too (Alberta, Canada, Australian package bees, Lounsberry *et al.*, 2010). Survival of adult and/or immature SHB life stages obviously depends on both storage conditions during transport and food availability. Moreover, inspections prior to and after trade (e.g. border controls) should also be considered. However, rather unlikely routes such as processed wax (Manitoba, Canada, cf. Neumann and Elzen, 2004) have also been reported. Genetic tools enable us to trace back the origin of invasive populations, which can be helpful to better mitigate future introductions. SHB mt-DNA sequence analyses of from the USA and South Africa indicate that the populations on both continents belong to a single species, although it is not clear whether a single or multiple introductions occurred (Evans *et al.*, 2000, 2003). In any case, the initial North American beetles shared the same source (Evans *et al.*, 2008; Lounsberry *et al.*, 2010). The outbreaks in Quebec, Canada, appear to originate from the USA (Evans *et al.*, 2003, 2008) and all have been found close to the USA border.

Discussion

The introduction of *A. tumida* in areas as far from its endemic range as North America, Australia, Europe and Asia illustrates the high potential of this beetle to spread around the globe (see Chapter Four). It is plausible that the import of package bees, honey bee and bumble bee colonies, queens, hive equipment and or even soil (Brown *et al.*, 2002) constitute potential invasion pathways of the SHB. Nevertheless, at the current state of evidence it is still unclear how the SHB actually reached the USA. The migratory nature of beekeeping is probably the greatest contributor of SHB transmission within in the USA (Neumann and Elzen, 2004). The pattern of SHB spread is probably dominated by long-distance jump dispersal as in other invasive species (Nentwig, 2007).

Since the 1996 introduction into the USA, SHB have become a global threat to both apiculture and wild bee populations. Despite comprehensive efforts, they will continue to spread. Their future impact will probably be more severe in areas with managed and feral European-derived honey bee populations as well as warm and humid climate, both fostering SHB population build-up (see Chapter Four). Therefore, particular concern should be addressed to beekeeping in Central and South America.

Beekeepers can adapt to SHB, and simple changes in management can go a long way to limit the impact of the beetle on managed bees; native bees are left to their own defences. In the early stages of SHB infestations in the USA, the average beekeeper might use chemical control around the honey house and even in the soil around the hives. Also the use of coumaphos traps was common within the hives. Currently, there are a number of non-chemical in-hive traps in wide use in the USA and coupled with good sanitation in both the honey house and the apiary have allowed beekeepers to manage SHB. This said, when bee colonies are under stress from *V. destructor* or others maladies and beetles can get a foothold in colonies,

SHB population explosions can occur. Lastly, queen breeders must constantly battle to keep SHB at low levels with traps in mating nucleus colonies and good sanitation measures in larger colonies. SHBs are well established in North America and continue their spread to South America. Beekeepers should be vigilant to their presence and learn from the mistakes of the past, as beetles ravaged honey houses and apiaries alike in the southern USA before management practices and better control options were adopted.

Acknowledgements

We wish to thank Pierre Giovenazzo and Hugh Smith for updates from Canada and Jamaica.

References

Annand, N. (2011). Investigations of small hive beetle biology to develop better control options. MSc thesis, University of Western Sydney, Australia.

Arbogast, R. T., Torto, B., Teal, P. E. (2010). Potential for population growth of the small hive beetle *Aethina tumida* (Coleoptera: Nitidulidae) on diets of pollen dough and oranges. *Florida Entomologist*, 93(2), 224-230.

Arias, H. D. M. (2014). Small hive beetle infestation (*Aethina tumida*), El Salvador. OIE report. http://www.oie.int/wahis_2/public/wahid.php/ Reviewreport/Review? page_refer=MapFullEventReport&reportid=14907

Berry, J. (2009). Small hive beetle round-up / Beetles come on strong in the south right now-be ready! *Bee Culture*, 137(10), 38-40.

Borroto, H., Chan, S. & Demedio, J. (2014). Diagnóstico de *Aethina tumida* Murray (Coleoptera: Nitidulidae) en colmenas (*Apis mellifera* L.) de Mayabeque, Memorias Jornadas Científicas por el 122 Aniversario del Sabio de la Medicina Veterinaria Cubana Dr. Ildefonso Pérez Vigueras, Universidad de Ciencias Médicas - Consejo Científico Veterinario. Pinar del Río, Cuba.

Brown, M. A., Thompson, H. M. & Bew, M. H. (2002). Risks to UK beekeeping from the parasitic mite *Tropilaelaps clareae* and the small hive beetle, *Aethina tumida*. *Bee World*, 83(4), 151-164. http://dx.doi.org/10.1080/0005772X.2002.11099558

Calderón Fallas, R. A., Montero, M. R., Arias, F. R., Villagra, W. V. (2015). Primer reporte de la presencia del pequeño escarabajo de la colmena *Aethina tumida*, en colmenas de abejas africanizadas en Nicaragua. *Cienc. Vet.* (in press).

Clay, H. (2006). Small hive beetle in Canada. *Hivelights*, 19, 14-16.

Connor, L. (2011a). Wipe out! The Big Island in crisis. *Bee Culture*, 139, 55-60.

Connor, L. (2011b), The Big Island in crisis: Part Two of the small hive beetle story in Hawaii. *Bee Culture*, 140, 23-27.

Darias, J. L. M. (2014). Small hive beetle infestation (*Aethina tumida*), Cuba. OIE report. http://www.oie.int/wahis_2/public/wahid.php/Reviewreport/Review?page_refer=MapFullEventReport&reportid=15673

de Guzman, L. I., Frake, A. & Rinderer, T. E. (2010). Seasonal population dynamics of small hive beetles, *Aethina tumida* Murray, in the south-eastern USA. *Journal of Apicultural Research*, 49(2), 186-191. http://dx.doi.org/10.3896/IBRA.1.49.2.07

Del Valle Molina, J. A. (2007). Small hive beetle infestation (*Aethina tumida*) in Mexico: Immediate notification report. Ref OIE: 6397, Report Date: 26/10/2007.

Dubuc, M. (2013). Small hive beetle infestation (*Aethina tumida*), Canada. OIE report. http://www.oie.int/wahis_2/public/wahid.php/Reviewreport/Review?page_refer=MapFullEventReport&reportid=14742

Ellis, J. D. & Hepburn, H. R. (2006). An ecological digest of the small hive beetle (*Aethina tumida*), a symbiont in honey bee colonies (*Apis mellifera*). *Insectes Sociaux*, 53 (1), 8-19.

Ellis, J. D. & Munn, P. A. (2005) The worldwide health status of honey bees. *Bee World*, 86(4), 88-101. http://dx.doi.org/10.1080/0005772X.2005.11417323

Evans, J. D., Pettis, J., Hood, M. W. M. & Shimanuki, H. (2003). Tracking an invasive honey bee pest: mitochondrial DNA variation in North American small hive beetles. *Apidologie*, 34, 103-109.

Evans, J. D., Pettis, J. & Shimanuki, H. (2000). Mitochondrial DNA relationships in an emergent pest of honey bees: *Aethina tumida* (Coleoptera: Nitidulidae) from the United States and Africa. *Annals of Entomological Society of America*, 93, 415-420.

Evans, J. D., Spiewok, S., Teixeira, E. W. & Neumann, P. (2008). Microsatellite loci for the small hive beetle, *Aethina tumida*, a nest parasite of honey bees. *Molecular Ecology Resources*, 8(3), 698-700.

Food and Environment Research Agency. (2010). *The small hive beetle: a serious threat to European apiculture*. Food and Environment Research Agency; Sand Hutton, UK. 23 pp.

Giovenazzo, P. & Bernier, M. (2015). Situation du petit coléoptère de la ruche au Québec. *L'Abeille*, 37(2), 8-9.

Giovenazzo, P. & Boucher, C. (2010). A scientific note on the occurrence of the small hive beetle (*Aethina tumida* Murray) in Southern Quebec. *American Bee Journal*, 150, 275-276.

Gordon, R., Bresolin-Schott, N. & East, I. J. (2014). Nomadic beekeeper movements create the potential for widespread disease in the honey bee industry. *Australian Veterinary Journal*, 92, 283–290.

Gutierrez, M. R. (2014). Small hive beetle infestation (*Aethina tumida*), Nicaragua. OIE report. http://www.oie.int/wahis_2/public/wahid.php/Reviewreport/Review page_refer=MapFullEventReport&reportid=14888

Hood, M. W. M. (2000). Overview of the small hive beetle, *Aethina tumida*, in North America. *Bee World*, 81(3), 129-137. http://dx.doi.org/10.1080/0005772X.2000.11099483

Hood, M. W. M. (2004). The small hive beetle, *Aethina tumida*: a review. *Bee World*, 85(3), 51-59. http://dx.doi.org/10.1080/0005772X.2004.11099624

Hood, M. W. M. (2011). Handbook of small hive beetle IPM. Clemson University, Cooperative Extension Service. Extension Bulletin 160, pp. 20. http://www.extension.org/sites/default/files/Handbook_of_Small_Hive_Beetle_IPM.pdf

Kozak, P. (2010). *Small hive beetle*. Ontario Ministry of Agriculture, Food and Rural Affairs; Guelph, Ontario, Canada. 4 pp.

Lounsberry, Z., Spiewok, S., Pernal, S. F., Sonstegard, T. S., Hood, M. W. M., Pettis, J., Neumann, P. & Evans, J. D. (2010). Worldwide diaspora of *Aethina tumida* (Coleoptera: Nitidulidae), a nest parasite of honey bees. *Annals of the Entomological Society of America*, 103(4), 671-677.

Loza, L. M. S., Álvarez, L. G. L., Ugalde, J. A. G. (2014). *Manual: Neuvos manejesos en la apicultura para el control del pequeño escarabajo de la colmena. SAGARPA*. http://www.sagarpa.gob.mx/ganaderia/Documents/final%20MANUAL%202da%20EDICI%C3%93N.pdf

Martin, S. J. (2013). Double trouble in paradise: small hive beetle joins varroa in Hawaii. *American Bee Journal*, 153(5), 529-532.

Milián, J. L. (2012). *Reporte de notificación de* Aethina tumida *a la OIE*. Dirección del Instituto de Medicina Veterinaria, Ministerio de la Agricultura, La Habana, Cuba.

Mustafa, S. G., Spiewok, S., Duncan, M., Spooner-Hart, R. & Rosenkranz, P. (2014). Susceptibility of small honey bee colonies to invasion by the small hive beetle, *Aethina tumida* (Coleoptera, Nitidulidae). *Journal of Applied Entomology*, 138(7), 547-550.

Nentwig, W. (2007). *Biological invasions*. Springer Verlag; Berlin, Germany.

Neumann, P. & Ellis, J. D. (2008). The small hive beetle (*Aethina tumida* Murray, Coleoptera: Nitidulidae): distribution, biology and control of an invasive species. *Journal of Apicultural Research*, 47(3), 181-183. http://dx.doi.org/10.3896/IBRA.1.47.3.01

Neumann, P. & Elzen, P. J. (2004). The biology of the small hive beetle (*Aethina tumida*, Coleoptera: Nitidulidae): Gaps in our knowledge of an invasive species. *Apidologie*, 35, 229-247.

Neumann, P; Hoffmann, D; Duncan, M; Spooner-Hart, R (2010) High and rapid infestation of isolated commercial honey bee colonies with small hive beetles in Australia. *Journal of Apicultural Research*, 49(4), 343-344. http://dx.doi.org/10.3896/IBRA.1.49.4.10

Neumann, P., Hoffmann, D., Duncan, M., Spooner-Hart, R. & Pettis, J. S. (2012). Long-range dispersal of small hive beetles. *Journal of Apicultural Research*, 51(2), 214-215. http://dx.doi.org/10.3896/IBRA.1.51.2.11

Pettis, J. S., Martin, D. & vanEngelsdorp, E. (2014). Migratory beekeeping. In *W. Ritter (Ed.), Bee Health and Veterinarians*. OIE, Paris, France. pp. 51-54.

Ramírez, M., Calderón, R. A., Hernández, P. & Benítez, J. (2014). Presencia del pequeño escarabajo de la colmena, Aethina tumida, en colmenas de abejas africanizadas en Centroamérica. *Bol. Parasitol.*, 15(3),1-2.

Robson, J. D. (2012). Small hive beetle Aethina tumida Murray (Coleoptera: Nitidulidae). Pest Alert 12-01. Plant Pest Control Branch, Division of Plant Industry, Department of Agriculture; Honolulu, Hawaii. http://hdoa.hawaii.gov/pi/files/2013/01/NPA-SHB-1-12.pdf

Spiewok, S. & Neumann, P. (2006). Cryptic low-level reproduction of small hive beetles in h o n e y   b e e colonies. *Journal of Apicultural Research*, 45(1), 47-48. http://dx.doi.org/10.1080/00218839.2006.11101313

Spiewok, S. & Neumann, P. (2012). Sex ratio and dispersal of small hive beetles. *Journal of Apicultural Research*, 51(2), 216-217. http://dx.doi.org/10.3896/IBRA.1.51.2.12

Spiewok, S., Duncan, M., Spooner-Hart, R., Pettis, J. S. & Neumann, P. (2008). Small hive beetle, Aethina tumida, populations II: Dispersal of small hive beetles. *Apidologie*, 39(6), 683-693.

Spiewok, S., Pettis, J. S., Duncan, M., Spooner-Hart, R., Westervelt, D. & Neumann P. (2007). Small hive beetle, Aethina tumida, populations I: Infestation levels of honey bee colonies, apiaries and regions. *Apidologie*, 38(6), 595-605.

Van Engelsdorp, D., Underwood, R., Caron, D. & Hayes, J. (2007). An estimate of managed colony losses in the winter of 2006 - 2007. A report commissioned by the apiary inspectors of America. *American Bee Journal*, 147, 599-603.

Jeff S Pettis[1], Marc O Schäfer[2] and Peter Neumann[3,4]

[1]USDA-ARS Bee Research Laboratory, Beltsville, Maryland, USA.

[2]National Reference Laboratory for Bee Diseases, Friedrich-Loeffler-Institute (FLI), Federal Research Institute for Animal Health, Greifswald Insel-Riems, Germany.

[3]Institute of Bee Health, Vetsuisse Faculty, University of Bern, Bern, Switzerland.

[4]Social Insect Research Group, Department of Zoology & Entomology, University of Pretoria, Pretoria, South Africa.

SIX *

A small hive beetle lesson from South Africa

Christian Pirk and Abdullahi Yusuf

Introduction

The recent detection of the small hive beetle (SHB) *Aethina tumida* (Figs 1-6) in Italy has triggered serious concerns within the apicultural and agricultural communities in Europe (Mutlinelli, 2014; Mutinelli *et al.*, 2014; Palmeri *et al.*, 2015). The previous chapters have dealt with their occurrence in Europe and the USA and the implications for different European conditions and legislations. As the chapters on SHB in new areas show, the beetle has the capacity to destroy healthy colonies of *Apis mellifera* of European origin. Although the SHB can be a serious threat to beekeeping operations when using European stock of *A. mellifera*, it is considered only a minor pest in its native range (Hepburn & Radloff, 1998).

The native range of the SHB is sub-Saharan Africa, which is characterised by varying degrees of beekeeping activities, with South Africa having an apicultural industry similar to Europe (Dietemann *et al.*, 2009). Although, SHB are seen as causes of colony losses in South Africa (Pirk *et al.*, 2014) the economic impact can be seen to be rather low (Johannsmeier, 2001). Beekeepers listing SHB as a potential

causes of colony loss did not lose more or fewer colonies than their counterparts, suggesting that the presence of SHB, although noticed by the beekeepers, is not triggering any kind of treatment (Pirk *et al.*, 2014).

Impact in South Africa

Depending on the region of South Africa, the impact of SHB is either slightly greater than that of the greater wax moth (*Galleria mellonella*) or slightly less; but significantly less than varroa or American foulbrood. It is therefore not threatening the beekeeping industry but rather is a nuisance, which can be dealt with by adopting specific management techniques. Moreover, until the expansion into north America in the 1990s (Neumann & Elzen, 2004; Neumann & Ellis, 2008; Chapter Five) only two publications

Fig. 1. Honey bee worker attacking (balling) an SHB (Photo: C. Laing).

Fig. 2. SHB running on honey and pollen comb. Notice that the honey bee workers in the foreground are ignoring the beetle (Photo: C. Laing).

addressing the biology or any aspects of *A. tumida* were published (Lundie, 1940). To date more than 200 peer-reviewed articles have addressed the introduction and effects of SHB in north America, Australia, north Africa and now Europe.

SHB can easily be noticed, even by inexperienced beekeepers when opening the hives, since beetles hide in cracks and cavities to which the workers have no access, or even run over the comb when disturbed (Figs 1-3,5). As soon as workers can reach the beetles they attack them and try to remove them from the colony (Neumann & Härtel, 2004; Neumann et al., 2001).

With the introductions of SHB into new ranges, bringing it into contact with honey bees of European origin, the question arises whether fundamental differences between the European and African subspecies of *Apis mellifera* might explain the huge differences in terms of impact the beetles had in North America (Eischen et al., 1998; Sanford, 1998) compared to within their native range.

We conducted behavioural experiments in Grahamstown, South Africa and Umatilla, Florida, USA, which revealed quantitative differences in the aggression levels of worker honey bees towards the beetle, but not qualitative ones (Elzen et al., 2001). Based on the quantitative differences one can conclude that the beetle is recognised as a threat by European honey bees. In addition, other behavioural patterns observed in African honey bees, such as prison building (Ellis et al., 2003b; Neumann et al., 2001) and being tricked into feeding the beetles (Ellis et al.,

Fig. 3. SHB being entombed in a crack in an *A. m. scutellata* colony (Photo: Christian W W Pirk).

2002b) (Fig. 4) were not only observed in African honey bees but also in European colonies (Ellis *et al.*, 2003b). The presence of SHB in colonies of European honey bees are said to trigger absconding behaviour (Ellis *et al.*, 2003d); non-reproductive swarming (Fig. 7) as a reaction to unfavourable conditions at the nesting site (Allsopp & Hepburn, 1997; Hepburn, 1988; Spiewok & Neumann, 2006; Spiewok *et al.*, 2006; Villa, 2004). The absconding behaviour was actually similar between Cape honey bees (*A. m. capensis*) in South Africa, and European honey bees in the USA when they are artificially infested with SHB (Ellis *et al.*, 2003e). These experiments were conducted for 15 days, and the colonies received 100 adult beetles each per day, and the amount of sealed brood, adult workers stored pollen and flight activity was recorded by Ellis and

Fig. 5. SHB trying to hide in cracks of the hive roof (Photo: C. Laing).

colleagues (2003e). The results show that European and Cape honey bees prepared slightly differentially for the coming absconding event (Fig. 7). The Cape honey bees had significantly reduced pollen stores. They kept the sealed brood area, the number of adults and the flight activity similar to the control colonies (no beetles added), whereas the European bees decrease sealed brood area, the number of adults and flight activity, but did not reduce pollen stores (*Ellis et al.*, 2003e). This would result in fewer workers defending the protein sources (pollen and brood) against SHBs. On the other hand the Cape workers did carry on as usual, thereby ensuring that enough workers are around to deal with the beetles. A reduced worker force results in pollen stores being more easily accessible

Fig. 4. Two SHB begging a honey bee worker for food (Photo: Christian W W Pirk).

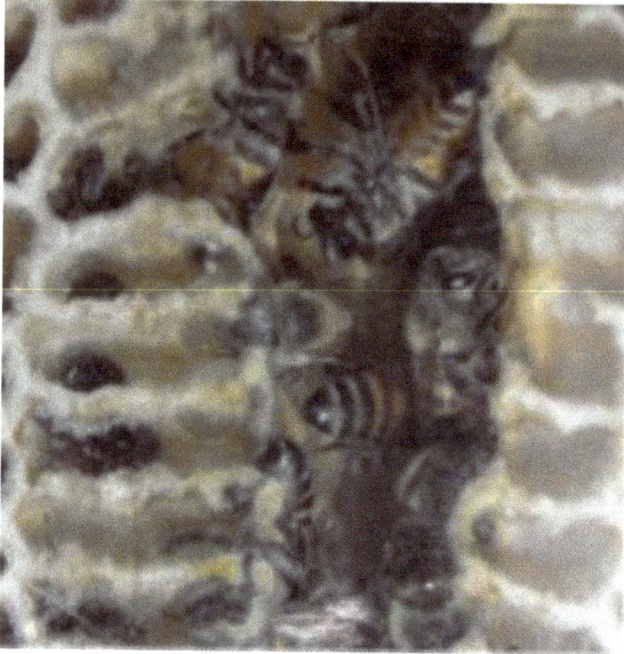

Fig. 6. SHB guarded by bees hiding at the bottom of empty cells (Photo: Christian W W Pirk).

by SHB due to fewer workers being around to protect the stores, which in itself could result in the following positive feedback loop. It has also been suggested that SHB facilitate the spread of a yeast (*Kodamaea ohmeri*) onto pollen, which in turn results in volatiles being realised from the inoculated pollen which attract more SHB (Torto *et al.*, 2007). However, the interactions between yeast, beetle and pollen stores are not limited to the invasive range of the beetle but also to the native ones, showing that this cannot explain the differences in the susceptibility of the different honey bee populations.

If it is not thus the subspecies *per se*, what can explain these clearly marked differences in the handling of SHB by their honey bee hosts? A recent study (Pirk & Neumann, 2013) has investigated the level of activity of honey bee workers and how that affects interactions with SHB. Young (<24 hours) and old workers (<7 days) were tested in their interactions with mature adult beetles. The results clearly show that the activity levels of young and older workers did not differ, so do not seem to play a significant role in how the beetle is sealed. Old workers attacked significantly more often than young workers, which in turn fed the beetles significantly more often (Pirk & Neumann, 2013). The level of aggressiveness or activity might influence the outcomes of the interactions between honey bees and SHB. If aggressiveness or activity is low, the beetles are ignored and fed by the hosts, which will mean that the host bees might fall prey to the beetle. The parasite will be kept under control if aggressiveness or activity is sufficiently high resulting in enough bees being around to chase the beetles into prisons or out of the hive.

Beekeeping practice

It seems therefore that there are only quantitative and not qualitative differences between the subspecies and that the activity levels might play a crucial role. The lesson to be learnt from the native range of SHB is that beekeeping is possible with only minor adjustments of beekeeping practice. In general, good beekeeping practice should be followed so that no sources of hive products are available for the small hive beetle to reproduce and continue its life cycle in. Good sanitation in the honey house, apiary and storage rooms are therefore necessary. During honey harvest, the beekeeper should either process the combs immediately or store the combs at 4°C or in sealed SHB proof containers. The same applies for old combs if storing combs is necessary; otherwise, one should refrain from doing so if possible. Bees should be supported in dealing with the beetle by ensuring that hive boxes have no or few cracks and cavities which allow the beetles to hide from the workers. It should be ensured that all parts of the hive are accessible by the workers. A particular problem is frames which are too close together so that workers cannot access the prime areas where the beetle will reproduce.

Fig. 7. Two swarms of *A. m. scutellata* settled near the old agricultural building of the University of Pretoria (Photo: Christian W W Pirk).

The space within the colony should correspond to the size of the honey bee population, thus ensuring that there are enough workers around to keep the SHB in check (Fig. 6). In particular, when the brood nest and/or the worker population is shrinking during winter or the dry season, beekeepers should either reduce the space available or supply workers, so that the worker force can cover the whole hive. Moreover, reducing the entrance of the colonies will probably assist them to deal with beetles before they even enter the colony (Ellis *et al.*, 2002a, 2003a; Neumann *et al.*, 2013).

SHB are a less important problem in the northern parts of South Africa than wax moth (Strauss *et al.*, 2013) and dealing with it is only as difficult as dealing with wax moth; one has to follow good beekeeping practice. In case of mass reproduction of SHB in colonies, it is important to remove all infested parts of the colony, e.g. combs. Since in such circumstances pupation of the SHB larvae will also be taking place in the soil outside the hives, treatment of the soil might be applicable to ensure that newly hatched beetles do not re-infest the apiary (Lundie, 1940; Neumann & Ellis, 2008; Neumann & Elzen, 2004). Outside the native range it seems also important to keep other parasites such as varroa under control, and one can use traps to reduce the beetle loads with individual colonies. See the other chapters for more information.

The rule of thumb is thus to help the bees to help themselves, by providing strong colonies, by providing them access to all parts of the hive where beetles could hide and/or reproduce. It is vital to not allow the beetle to reproduce outside the colony, such as in the honey house or in stored old comb. These adjustments of beekeeping practice can make it relatively easy to deal with the presence of SHB.

References

Allsopp, M. H., & Hepburn, H. R. (1997). Swarming, supersedure and mating system of a natural population of honey bees (*Apis mellifera capensis*). *Journal of Apicultural Research, 36(1)*, 41-48.
http://dx.doi.org/10.1080/00218839.1997.11100929

Dietemann, V., Pirk, C. W. W., & Crewe, R. M. (2009). Is there a need for conservation of honey bees in Africa? *Apidologie, 40*, 285-295.

Eischen, F. A., Baxter, J. R., Elzen, P. J., Westervelt, D., & Wilson, W. T. (1998). Is the small hive beetle a serious pest of US honey bees? *American Bee Journal, 138*(12), 882-883.

Ellis, J. D., Delaplane, K. S., Hepburn, R., & Elzen, P. J. (2002a). Controlling small hive beetles (*Aethina tumida* Murray) in honey bee (*Apis mellifera*) colonies using a modified hive entrance. *American Bee Journal, 142(4),* 288-290.

Ellis, J. D., Delaplane, K. S., Hepburn, R., & Elzen, P. J. (2003a). Efficacy of modified hive entrances and a bottom screen device for controlling *Aethina tumida* (Coleoptera : Nitidulidae) infestations in *Apis mellifera* (Hymenoptera : Apidae) colonies. *Journal of Economic Entomology, 96(6),* 1647-1652.
http://dx.doi.org/10.1603/0022-0493-96.6.1647

Ellis, J. D., Hepburn, H. R., Ellis, A. M., & Elzen, P. J. (2003b). Social encapsulation of the small hive beetle (*Aethina tumida* Murray) by European honey bees (*Apis mellifera* L.). *Insectes Sociaux, 50(3),* 286-291.
http://dx.doi.org/10.1007/S00040-003-0671-7

Ellis, J. D., Hepburn, R., Delaplane, K. S., & Elzen, P. J. (2003d). A scientific note on small hive beetle (*Aethina tumida*) oviposition and behaviour during European (*Apis mellifera*) honey bee clustering and absconding events. *Journal of Apicultural Research, 42(3),* 47-48.
http://dx.doi.org/10.1080/00218839.2003.11101089

Ellis, J. D., Hepburn, R., Delaplane, K. S., Neumann, P., & Elzen, P. J. (2003e). The effects of adult small hive beetles, *Aethina tumida* (Coleoptera : Nitidulidae), on nests and flight activity of Cape and European honey bees (*Apis mellifera*). *Apidologie, 34(4),* 399-408.
http://dx.doi.org/10.1051/apido:2003038

Ellis, J. D., Pirk, C. W. W., Hepburn, H. R., Kastberger, G., & Elzen, P. J. (2002b). Small hive beetles survive in honey bee prisons by behavioural mimicry. *Naturwissenschaften, 89,* 326-328.

Elzen, P. J., Baxter, J. R., Neumann, P., Solbrig, A., Pirk, C., Hepburn, H. R., Westervelt, D. & Randall, C. (2001). Behaviour of African and European subspecies o f *Apis mellifera* toward the small hive beetle, *Aethina tumida. Journal of Apicultural Research, 40(1),* 40-41.
http://dx.doi.org/10.1080/00218839.2001.11101049

Hepburn, H. R. (1988). Absconding in the African honey bee - the queen, engorgement and wax secretion. *Journal of Apicultural Research, 27(2),* 95-102. http://dx.doi.org/ 10.1080/00218839.1988.11100787

Hepburn, H. R., & Radloff, S. E. (1998). *Honey bees of Africa.* Springer Verlag; Berlin, Germany.

Johannsmeier, M. F. (2001). *Beekeeping in South Africa:* ARC-Plant Protection Research Institute; South Africa.

Lundie, A. E. (1940). The small hive beetle, *Aethina tumida. Science Bulletin Union of South Africa, 220,* 5-19.

Mutinelli, F. (2014). The 2014 outbreak of the small hive beetle in Italy. *Bee World, 91(4),* 88-89. http://dx.doi.org/10.1080/0005772X.2014.11417618

Mutinelli, F., Montarsi, F., Federico, G., Granato, A., Maroni Ponti, A., Grandinetti, G., Ferrè, N., Franco, S., Duquesne, V., Rivière, M.-P., Thiéry, R., Henrikx, P., Ribière-Chabert, M. & Chauzat, M.-P. (2014). Detection of *Aethina tumida* Murray (*Coleoptera: Nitidulidae.*) in Italy: outbreaks and early reaction measures. *Journal of Apicultural Research, 53,* 569-575. http://dx.doi.org/10.3896/IBRA.1.53.5.08

Neumann, P. & Ellis, J. D. (2008). The small hive beetle (*Aethina tumida* Murray, Coleoptera: Nitidulidae): distribution, biology and control of an invasive species. *Journal of Apicultural Research, 47(3),* 181-183. http://dx.doi.org/10.3896/IBRA.1.47.3.01

Neumann, P., & Elzen, P. J. (2004). The biology of the small hive beetle (*Aethina tumida,* Coleoptera : Nitidulidae): Gaps in our knowledge of an invasive species. *Apidologie, 35(3),* 229-247.

Neumann, P., Evans, J. D., Pettis, J. S., Pirk, C. W. W., Schäfer, M. O., Tanner, G. & Ellis, J. D. (2013). Standard methods for small hive beetle research. In *V. Dietemann, J. D. Ellis & P. Neumann (Eds) The COLOSS BEEBOOK: Volume II: Standard methods for* Apis mellifera *pest and pathogen research. Journal of Apicultural Research, 52(4),* http://dx.doi.org/10.3896/IBRA.1.52.4.19

Neumann, P., & Härtel, S. (2004). Removal of small hive beetle (*Aethina tumida*) eggs and larvae by African honey bee colonies (*Apis mellifera scutellata*). *Apidologie, 35(1),* 31-36.

Neumann, P., Pirk, C. W. W., Hepburn, H. R., Solbrig, A. J., Ratnieks, F. L. W., Elzen, P. J., & Baxter, J. R. (2001). Social encapsulation of beetle parasites by Cape honey bee colonies (*Apis mellifera capensis* Esch.). *Naturwissenschaften, 88(5),* 214-216. http://dx.doi.org/10.1007/s001140100224

Palmeri, V., Scirtò, G., Malacrinò, A., Laudani, F. & Campolo, O. (2015). A new pest for European honey bees: first report of *Aethina tumida* Murray (Coleoptera Nitidulidae) in Europe. *Apidologie, 46(4),* 527-529. http://dx.doi.org/10.1007/s13592-014-0343-9

Pirk, C. W. W., Human, H., Crewe, R. M., & vanEngelsdorp, D. (2014). A survey of managed honey bee colony losses in the Republic of South Africa - 2009 to 2011. *Journal of Apicultural Research, 53(1)*, 35-42. http://dx.doi.org/10.3896/IBRA.1.53.1.03

Pirk, C. W. W., & Neumann, P. (2013). Small hive beetles are facultative predators of adult honey bees. *Journal of Insect Behavior, 26*, 796-803. http://dx.doi.org/10.1007/s10905-013-9392-6

Sanford, M. T. (1998). *Aethina tumida*: a new bee hive pest in the Western Hemisphere. *APIS (University of Florida), 16 (7)*, 1-5.

Spiewok, S., & Neumann, P. (2006). The impact of recent queenloss and colony phenotype on the removal of small hive beetle (*Aethina tumida* Murray) eggs and larvae by African honey bee colonies (*Apis mellifera capensis* Esch.). *Journal of Insect Behavior, 19(5)*, 601-611. http://dx.doi.org/10.1007/S10905-006-9046-Z

Spiewok, S., Neumann, P., & Hepburn, H. R. (2006). Preparation for disturbance-induced absconding of Cape honey bee colonies (*Apis mellifera capensis* Esch.). *Insectes Sociaux, 53*, 27-31.

Strauss, U., Human, H., Gauthier, L., Crewe, R. M., Dietemann, V., & Pirk, C. W. W. (2013). Seasonal prevalence of pathogens and parasites in the savannah honey bee (*Apis mellifera scutellata*). *Journal of Invertebrate Pathology, 114(1)*, 45-52. http://dx.doi.org/10.1016/j.jip.2013.05.003

Torto, B., Boucias, D. G., Arbogast, R. T., Tumlinson, J. H., & Teal, P. E. A. (2007). Multitrophic interaction facilitates parasite-host relationship between an invasive beetle and the honey bee. *Proceedings of the National Academy of Sciences of the United States of America, 104 (20)*, 8374-8378.

Villa, J. D. (2004). Swarming behavior of honey bees (Hymenoptera: Apidae) in southeastern Louisiana. *Annals of the Entomological Society of America, 97*, 111-116.

Christian W W Pirk and Abdullahi A Yusuf

Social Insect research Group, Department of Zoology & Entomology, University of Pretoria, Private Bag X20, Hatfield 0028, Pretoria, South Africa.

Email: cwwpirk@zoology.up.ac.za

SEVEN

The small hive beetle in Australia

Robert Spooner-Hart, Nicholas Annand and Michael Duncan

Introduction

There is some discrepancy among reports of the initial discovery of the small hive beetle (SHB), *Aethina tumida* (Murray) in Australia. Nevertheless, it is undisputed that beetles were first observed in hives at Richmond, New South Wales (NSW) (33.5653°S, 150.7597°E) and at the University of Western Sydney's (UWS) Hawkesbury Campus at East Richmond (33.6206°S, 150.7317°E) 2 km distant, and that beetle samples were submitted for identification to NSW Department of Agriculture in early July 2002. The beetles were sent to the Orange Agricultural Institute and were processed and reported in a routine manner as they were thought to be the same as an undescribed endemic Australian species found on the South Coast during 2001, and in late July were identified as *Aethina* sp. (Spence, 2002). However, it was not until 24 October that the specimens were sent to CSIRO Division of Entomology where they were identified as *A. tumida* on 25 October. A confirmatory diagnosis was provided by a coleopterist at the Australian National Insect Collection on 31 October. At this time, beetles were also reported to be in hives at Camperdown (33.8880°S, 151.1869°E) 50 km distant, Gosford (33.4267°S, 151.3417°E) 50 km direct and 90 km by road, and Wedderburn (34.1261°S, 150.8179°E) 70 km (Spence 2002), presumably through movement of infested hives.

Initial response to the incursion

At the time of its discovery, SHB was a declared disease under the NSW Exotic Diseases of Animals Act, but not covered by the compensation provisions of the Act. It was also a Category 3 disease under the Emergency Animal Disease Response Agreement (50:50 industry: government funding), but there was no specific AUSVETPLAN strategy (planned national, state and district level responses to animal disease emergencies) for this pest (Fogarty 2002) although a general strategy was in place for exotic bee diseases and pests.

On the day of the original diagnosis (25 October 2002), the UWS and Richmond apiaries were quarantined and tracing, inspection and quarantining of contact apiaries commenced. A Section 76 Declaration of Exotic Disease was issued on 1 November 2002, and a Restricted Area (RA) was implemented preventing the movement of any bees or bee products within, into or outside the area covering the Sydney metropolitan area and Gosford, with 33 properties quarantined (Toffolon, 2002). A state disease control headquarters (SDCHQ) was established on 31 October 2002, and a local disease control centre (LDCC) was subsequently established at UWS on 5 November 2002. The SDCHQ continued to manage those areas of the state outside the Sydney Basin RA.

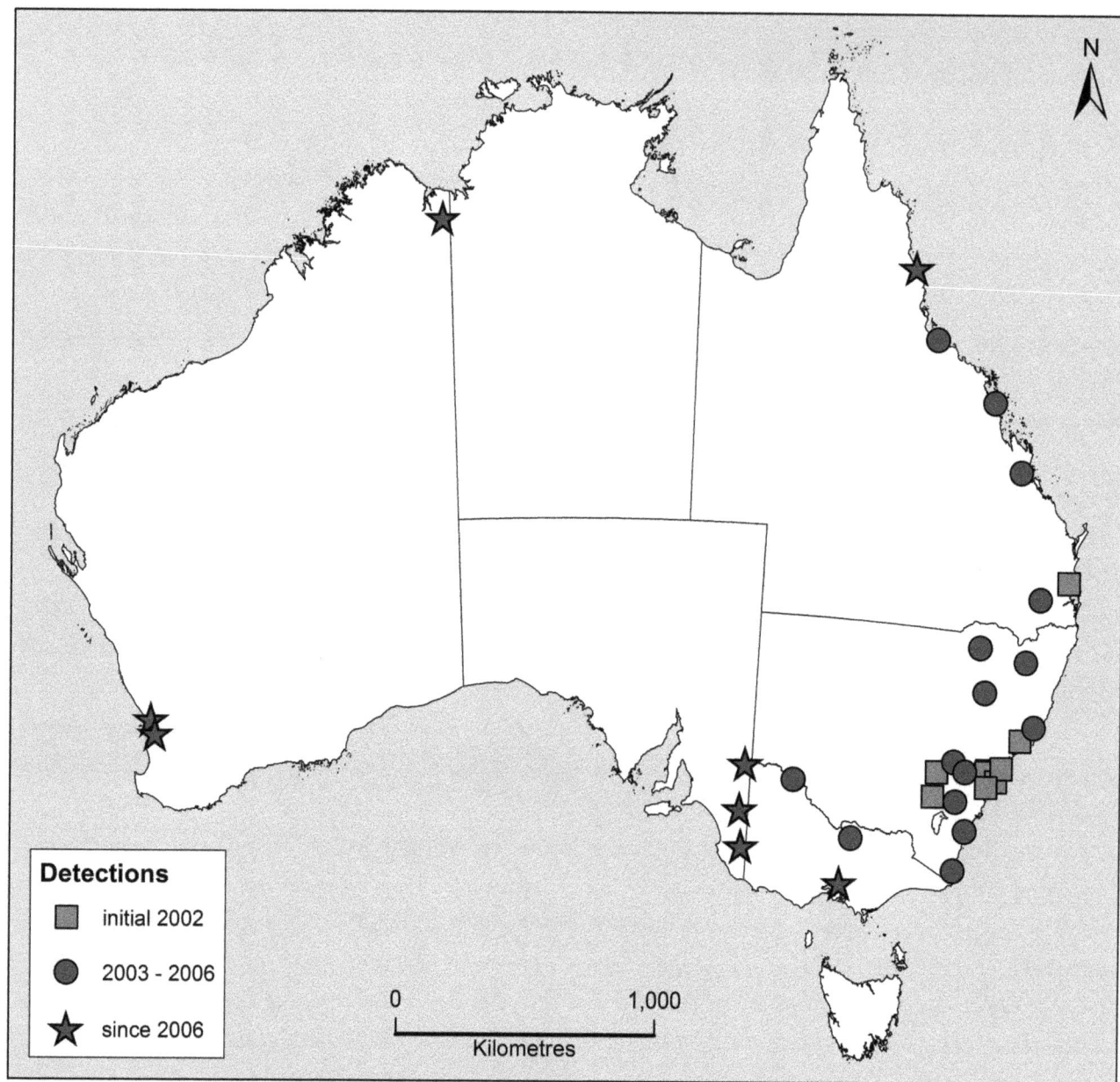

Fig. 1. Map showing detections of small hive beetle in Australia.

Subsequently, RAs were established at Stroud (32.4036°S, 151.9670°E) (5 November), Cowra (33.8350°S, 148.6925° E) (7 November) and Binalong (34.6667°S, 148.6500°E) (7 November), all associated with hive movements from the Sydney Basin (Spence, 2002). A Small Hive Beetle National Management Group was established and, at its direction, Animal Health Australia formed a SHB Steering Committee (Animal Health Australia, 2003).

A survey was initially conducted under the supervision of the SDCHQ. The first phase of the survey was to ascertain the extent of the incursion. Regulatory staff from NSW Agriculture examined the history of movements of infested hives and scrutinised any hives associated with these movements (Gillespie et al., 2003). All commercial and feral hives (where identified) within 3 km of a confirmed infestation were also inspected and commercial apiaries up to 10 km could be included in this search, dependent on environment and location of these commercial apiaries (Bell undated; Bell, cited in Gillespie et al., 2003). However, it was soon concluded that that the eradication of SHB infestation was not feasible, because of its spread as well as its presence in feral colonies and on 26 November 2002 the National Emergency Animal Disease Management Group announced that the eradication of SHB infestation in Australia would not be attempted, and a national strategy would be developed to assist beekeepers to manage SHB infestations (Spence, 2002).

In the second phase of the survey, all 3,200 registered beekeepers throughout NSW were mailed on 2 December 2002 with advice of the SHB problem and a survey form, with a request to inspect their hives and submit samples of any beetles found. This survey database was interrogated in January 2003; of the 1,059 reports on the database, there were 120 positive detections with 12 of these in feral hives (Gillespie et al., 2003). Of these detections, three infected apiaries were in the Stroud RA, 7 in the Cowra RA (including one feral hive) and 1 in the Binalong RA (Gillespie et al., 2003).

Given the degree of spread and establishment, Gillespie et al. (2003) concluded that SHB was likely to have been in NSW for at least six months prior to its discovery. Our view, consistent with that of Animal Health Australia (2003), is that this is an underestimate and it is likely that SHB had been present for more than one year.

The subsequent distribution and status of SHB has been reported by Rhodes and McCorkell (2007) in NSW following a survey in 2006, and by Neumann & Ellis (2008) and Annand (2011). Interviews were conducted by one of the authors (N. A.) in May 2015 with state government apiculture officers to update this report. The following is a compilation of all these data. A map showing the reported SHB detections in Australia is presented as Fig. 1. The detections are separated into 3 phases: 2002, the initial detections; 2003-2006, subsequent detections showing spread associated with movement of hives soon after the initial detections; and post-2006, the most recent detections showing spread to non-eastern Australian states.

Table 1. List of small hive beetle projects funded by Rural Industries Research & Development Corporation.

Years	Project Title	Organisation responsible	Comment/ Publications
2003	Study of the small hive beetle in the USA	NSW Department of Agriculture	Somerville (2003)
2003-2004	Using temperature manipulation to control small hive beetle	NSW Department of Agriculture	
2003-2004	Insecticidal control of small hive beetle Part 1	NSW Department of Agriculture	Levot (2009)
2003	Investigate small hive beetle traps	NSW Department of Agriculture	No report. Probably rolled into insecticidal control project
2004	Investigating attractants and pheromones for small hive beetle management	Victorian Department of Primary Industries	Not funded due in part to funding shortfall
2004-2007	Insecticidal control of small hive beetle Part 2	NSW Department of Agriculture	Levot (2007, 2009)
2006-2010	Small hive beetle biology providing options	NSW Department of Agriculture	Annand (2011a, 2011b)
2006-2007	Sustainable control of small hive beetle through targeting in-ground stages	UWS	Spooner-Hart (2008)
2007-2008	Feasibility study into in-hive fungal bio-control of small hive beetle	Queensland DPI	Leemon (2009)
2008-2009	Evaluation of anti-varroa boards for increase in honey production (included small hive beetle)	UWS	Spooner-Hart (2010); Keshlaf & Spooner-Hart (2013)
2009-2011	Commercialisation of the small hive beetle harbourage device	NSW Department of Industry & Investment*	Levot (2012)
2010-2011	In hive fungal bio-control of small hive beetle	Queensland DPI	Leemon (2012)
2013-2014	APITHOR™ small hive beetle harbourage trap safety and residue trials	NSW Department of Primary Industries*	Levot (2014)
2014-2017	External attractant trap for small hive beetle	Queensland DPI	

SHB in New South Wales and ACT

Following the initial detections, SHB spread rapidly throughout much of eastern NSW, primarily through movement of infested hives. By mid-autumn 2003, SHB was found along much of the coast from Sydney north to Taree (31.9000°S, 152.4500°E) (Rankmore, 2003; 2007, cited in Annand, 2011) a distance of approx. 300 km. By the end of 2003, SHB was found from Batemans Bay (35.7081°S, 150.1744°E) in the south to Vittoria (33.4331°S, 149.3515°E, esl 981 m), Gunnedah (30.9667°S, 150.2500°E), Moree (29.4658°S, 149.8339°E) and Glen Innes (29.7500°S, 151.7361°E, esl 1062 m) to the west and north, a distance north-south of more than 800 km (Annand 2011). SHB has since continued to spread throughout the state (Rankmore, 2003; 2007, cited in Annand, 2011). An indication of the rate of spread is that of the 15 districts surveyed by Rhodes and McCorkell in NSW, SHB was present in three districts in 2002, six districts in 2003, 13 districts in 2004 and all 15 districts in 2006. SHB numbers and damage to hives have been worst on the eastern side of the Great Dividing Range north of Batemans Bay, where humid, hot conditions are more prevalent (Rhodes and McCorkell, 2007). In the dry, far western areas, the cool tablelands and the south of the state SHB has been far less troublesome (Rhodes and McCorkell, 2007), probably due to less favourable climatic conditions for completion of its life cycle. However, a change in conditions to a much wetter spring and summer in 2010-11 resulted in beekeepers reporting hive losses in areas previously thought of as low risk to SHB attack.

Reports of SHB damage from the south coastal town of Eden (37.0667°S, 149.9000°E) close to the Victorian border and on the cooler ranges including the Blue Mountains (Clinton, 2011, cited in Annand, 2011), Goulburn (34.7547°S, 149.6186°E, 702 m esl) (Somerville, 2011, cited in Annand, 2011) and Oberon (33.7167°S, 149.8667°E, 1100 m esl) (Taylor, 2011, cited in Annand, 2011) suggest that it can be a major pest through most of the state when conditions are favourable. In some areas it is hard to ascertain whether SHB is established, or if beetles only accompany hive movements. For the purpose of SHB spread, the Australian Capital Territory (ACT) is considered as part of NSW. The tableland climate of the ACT appears to limit SHB establishment as a major pest; however, there have been anecdotal reports of SHB-caused hive losses in the summer of 2010-11 (Annand, 2011).

Prior to the arrival of SHB, most eastern Australian queen rearing activities were conducted from the Sydney basin northwards, along the coast to southern Queensland. (Qld) Apart from the direct damage to hives, especially nucleus hives caused by SHB, the quarantine restrictions imposed nationally and internationally, severely impacted domestic and export sales of queens reared in SHB-declared areas. As a result, many large queen breeders either ceased business or moved their operations over the Great Dividing Range to where conditions are cooler and less humid. This has proved to be an effective SHB management strategy. Currently, the major SHB activity remains on central and north coastal NSW with continuing severe SHB pressure in the Sydney basin. We have recently observed that even apparently strong, healthy hives can be severely impacted by SHB.

SHB in Queensland

The Queensland government was notified on the 28 October 2002 of hive movement from the Richmond, NSW district and SHB was found in hives in Beerwah (26.8556°S, 152.9600°E) in coastal Qld and six apiary sites in close proximity also contained SHB (Lamb, 2007, cited in Annand, 2011). In April 2003, apiary inspectors conducted hive inspections for SHB at 40 locations scattered throughout the state and found no further infestations. However, in May 2003 an inspection of hives within a 15 km radius around the initial infested colonies showed that SHB was naturally spreading out and inhabiting hives in the locality (Lamb, 2007, cited in Annand, 2011). A mail-out survey was sent to all registered beekeepers in Qld in 2003, to which 98 replied with 3 reporting they had seen SHB and with one positive sample returned (Lamb, 2007, cited in Annand, 2011).

After this date, the rate of spread escalated, no doubt through the movement of hives. SHB was found in Toowoomba (27.5667°S, 151.9500°E) in October 2004, by January 2006 was in Rockhampton (23.3750°S, 150.5117°E) and by June 2006 in Mackay (21.1411°S, 149.1861°E) and Townsville (19.2564°S, 146.8183°E) (Lamb, 2007, cited in Annand, 2011). Similar to NSW, SHB has caused more problems to beekeepers along the more humid coastal areas of Qld. SHB was found in Cairns (16.9256°S, 145.7753°E) and the Atherton Tablelands in northern Qld in 2011. Interestingly, it was also found at the same time in *Apis cerana* Java strain colonies in the Cairns area following an established 2007 incursion. Currently, the main distribution of SHB is along the coast and inland to Goondiwindi in south-east Qld. There is no continuing formal monitoring or reporting (Hamish Lamb, *pers. comm.* May 2015)

SHB in Victoria

SHB was first detected in Victoria (Vic.) in August 2003 in hives of a NSW beekeeper (Joe Riodan, *pers. comm.* May 2015), and the hives were immediately returned. Subsequent incursions were detected in August 2005 in hives from the NSW south coast (Joe Reardon, *pers. comm.* May 2015) and again in August 2007, associated with the mass movement of colonies to pollinate almonds in the dry north-west corner of Vic. (Kaczynski, 2007, cited in Annand, 2011). SHB can now be regularly found across most of Vic., with the majority of hives containing some SHB. The Mediterranean summers combined with the dry conditions from 2003-2009 probably helped slow its establishment in Vic., whereas wet conditions in the 2010-11 summer resulted in SHB become more troublesome (Martin, 2011, cited in Annand, 2011). The areas where beekeepers have experienced problems with SHB and occasional hive break downs are the north-east, central and Melbourne districts (Riordan, 2010, cited in Annand, 2011). The fact that SHB has yet to become a well-established major pest for beekeepers in this state suggests that conditions are sub-optimal.

SHB is currently present throughout much of Vic., with numbers peaking during hot, humid conditions. It can be found in healthy strong honey bee colonies although without causing major damage, but particularly targets hives that are compromised/diseased/drone layers, and only seriously impacts colonies with poor beekeeper management practices (Joe Riordan, *pers. comm.* May 2015).

SHB in Western Australia

In September 2007 SHB was identified in beehives in the far north of Western Australia (WA) at Kununurra (15.7736°S, 128.7386°E) (Trend, 2010, cited Annand, 2011). The incursion is suspected to have been introduced via diseased bee boxes that had been sent to south-east Qld for irradiation in July 2006 and returned containing SHB (Manning, 2008, cited in Annand, 2011; Bill Trend, *pers. comm.* May 2015). Some hives moved from Kununurra were traced to southern WA where SHB was found in October 2007 at West Swan (31.8500°S, 115.9770°E) and Jarrahdale (32.3390°S, 116.0620°E), near Perth. These hives (62) were destroyed immediately and no further SHB has been detected in the southern half of the state (Bill Trend, *pers. comm.* May 2015). As a result of these incursions, quarantine legislation preventing movement of hives and

disinfection and inspection of bee equipment prior to its movement was introduced in 2008. A SHB population remains in Kununurra. Regular inspections and monitoring was conducted using oil reservoir traps as well as sentinel hive surveillance (approx. 150 hives) between 2008-2012 in south-west Western Australia (Bill Trend, *pers. comm.* May 2015).

SHB in South Australia

South Australia (SA) is the most recent state in Australia in which SHB is becoming established. There have been several detections of SHB in SA. The first two were from Vic. hives transported over the border, but more recent detections have been in local hives. The first was in May 2011 when several adult SHB were found in illegally moved hives at Ngarkat Conservation Park (35.7167°S, 140.6000°E) on the SA-Vic. border, and which were moved back to Vic. The second was in May 2012 at Naracoorte (36.9500°S, 140.7500°E) in eastern SA near the SA-Vic. border, where beetles and larvae were found in hives fed supplementary pollen. The third detection was in December 2014 in Renmark (34.1667°S, 140.7333°E) in the Riverland area. SHB continues to be found in apiaries around Renmark. In April 2015 a total of 62 SHB adults were collected from surveillance trapping from 9 separate small apiaries belonging to seven different beekeepers (Michael Stedman, *pers. comm.* May 2015).

SHB in Northern Territory

The Northern Territory (NT) detected an incursion of a single adult SHB in September 2010 in a queen bee shipment from Qld. Protocols for the entry of queens into the NT were subsequently changed to minimise the chance of a repeat incursion via queen introductions. Monitoring for SHB continues with the sentinel hives, and no further detections have been recorded (Vicki Simlesa, pers comm. May 2015).

SHB and Tasmania.

There have been no detections of SHB in Tasmania (Tas.), where sentinel hives are monitored at four air/sea ports for SHB (Karla Williams, pers. comm. May 2015). Queen bees entering Tas. require inspection both prior to transport and on arrival in Tas., prior to release (Andrewartha, 2013).

SHB associated with bee species other than *Apis mellifera*

As previously reported, SHB was recorded in colonies of *Apis cerana* in the Cairns area of Qld following an established incursion there, but is no longer monitored for.

Stingless bees

Australia is home to approximately 15 species of stingless bees in two genera *Tetragonula* and *Austroplebeia* (Halcroft et al., 2013), with six *Tetragonula* species (Michener, 2013) and the Australian *Austroplebeia* species being delineated (Halcroft et al., submitted) and Australian and New Guinean species under revision (Anne Dollin, *pers. comm.* May 2015). Stingless bees are essentially tropical, only thriving in warm areas of Australia such as Qld, northern areas of WA and NT, and north-eastern areas of NSW (Dollin, http://www.aussiebee.com.au/australian-stingless-bees.html#question2), except *T. carbonaria* which is distributed as far south as Bega NSW (36.6667°S). There are anecdotal reports of SHB in reasonably healthy stingless bee hives, most commonly after rehiving or splitting (e.g., Halcroft, 2012; Wade, 2012), as well as a report of them entering healthy hives during a greenhouse investigation), (Neumann et al 2012) although no SHB have been observed in *T. carbonaria* hives continuously maintained at UWS Richmond except those that have died-out through other causes. Greco *et al.* (2010) reported that workers of *T. carbonaria* immediately mummify invading adult small hive beetles alive by coating them with a mixture of resin, wax and mud and Halcroft *et al.* (2010) reported strong and rapid defensive behaviours in *A. australis* against introduced SHB eggs, 3-day-old larvae, and adult beetles (including entombing them), concluding that this species also has well developed hive defences likely to minimize entry and survival of SHB.

Beekeeper response to SHB incursion

There was initially a degree of anger from some beekeepers directed to the NSW Department of Agriculture (e.g. Malfroy, 2003) for the perceived lapses in surveillance and delayed identification of the SHB. Although initial reports indicated hive losses from SHB were low (Neumann & Elzen, 2004) it was not long before this situation changed. Following a 2006 NSW survey, Rhodes & McCorkell (2007) reported that more than 4500 colonies had been lost in that state since the incursion in 2002, with average estimated costs of A$10,529 per beekeeper over this time. Annand (2011) reported a number of commercial apiarists had lost up to 30% of their hives. SHB also caused serious problems for queen breeders and bee package producers located in the Sydney basin and central eastern Australian coast. A SHB incursion in Canada was attributed to importation of packaged bees from Australia (Lounsberry et al., 2010), resulting in closure of the packaged bee trade from eastern Australia into Canada in 2007 (Annand, 2011).

In the absence of registered pesticides, beekeepers were initially unable to legally use any chemical control measures for SHB in their apiaries. The granting in December 2002 of a permit by the Australian Pesticides and Veterinary Medicines Authority (APVMA) for use of permethrin as a soil drench around bee hives or for ground intended for hive placement provided beekeepers with a limited option for out of hive treatment of SHB. The APVMA subsequently (December 2007) granted a permit for aluminium phosphide to fumigate SHB-infested combs following a favourable report from trials by Levot & Haque (2006a). However, there was no legal option for beekeepers for in-hive control.

In the absence of commercially available traps, a number of beekeepers employed home-made traps (such as fashioned from compact disc covers or fishing tackle boxes, see Annand, 2011), in some cases incorporating insecticides not registered for use in hives. However, two commercial in-hive traps were developed by beekeepers. These were AJs Beetle Eater™ (a slotted trap containing vegetable oil in its base) (http://www.ajsbeetleeater.com.au/) which was shown to be effective in trials conducted in Canada (Bernier et al., 2015), and BEETLTRA, a black plastic tray and slide rails fitted externally underneath the beehive bottom board. The latter trap is recommended by its manufacturers for use in conjunction with drilled beetle escape holes in the hive bottom board (Heenan & Heenan, 2007) and with hydrated lime or fluid, or oil (Annand 2008) in the tray; beekeepers using diatomaceous earth have reported this combination to be highly efficacious (Tim Malfroy, pers. comm. June 2015). There is now also wide availability of the APITHOR™ trap developed by Levot (2007; 2008, 2009; 2012; 2014) as a single use, disposable in-hive device (https://www.apithor.com.au/), containing fipronil.

As previously reported, many larger queen breeders and bee package producers located in the Sydney Basin and the east coast were severely impacted by SHB, and either ceased business or moved their operations over the Great Dividing Range to where conditions are cooler and less humid. This has proved to be an effective SHB management strategy. Queen breeders still located in areas with SHB have moved away from use of mini-nucleus colonies in favour of larger, stronger nucleus colonies that are more capable of resisting SHB (Mustafa et al., 2014).

Most beekeeping in Australia is conducted in 8- or 10-frame Langstroth hives. There is increasing interest in Warré beekeeping, particularly by hobbyist beekeepers, with natural comb and reduced hive management/manipulation practices, and there are reports that while adult SHB may be present in low numbers in these hives, there have been no significant infestations with reproduction (Tim Malfroy, pers. comm. June, 2015).

Funded SHB research projects

Following the decision that eradication of SHB infestation was not feasible and that a national strategy would be developed to assist beekeepers to manage SHB infestations, Animal Health Australia was requested to coordinate and broker funding for a National SHB Management Plan (Animal Health 2003). The objectives of the plan were to: 1. Reduce the impact on productivity, slow the spread of SHB in Australia and minimise the damage in infected apiaries by identifying chemical, non-chemical and management measures; 2. Implement cost-effective surveillance to enable the spread to be monitored and reported; 3. Develop an on-going communications program to keep beekeepers and horticultural industries informed; and 4. Provide cost effective national coordination and review.

The participants of SHB National Management Plan together with the SHB NMG and the Rural Industries Research & Development Corporation (RIRDC) initiated a SHB Expert Steering Committee comprising representatives of the Commonwealth and affected State governments, RIRDC and the Australian Honey Bee Industry Council. A number of key research projects and activities were identified and funded as part of the National Management Plan response for the period 2002-2004 with an operating budget of A\$177,000 and A\$867,000 in-kind. The first activity was a fact-finding study tour by government and industry representatives to USA in March 2003, visiting a wide range of research institutions and beekeepers (Somerville, 2003). Other projects included investigating in-hive and in-ground chemical treatments, in-hive traps, chilling and freezing and SHB attractants. The management plan response also included a surveillance, monitoring and reporting system, managed at state level.

A list of all the SHB projects funded through the Honey bee (now the Honey bee and Pollination) program of RIRDC, established in 1989 by the Australian Government to work with industry to invest in research and development, is provided in Table 1. In addition to work funded by RIRDC, a significant contribution to SHB research in Australia and internationally was made by Prof. Peter Neumann, University of Bern, Switzerland, (formerly Martin-Luther University, Halle, Germany) and his team, including Sebastian Spiewok, Sven Buchholz and Dorothee Hoffmann and Sandra Mustafa who visited Australia four times for research activities from summer 2005 to autumn 2007, as part of a UWS-funded visiting international research initiative scheme. For Neumann, this was an extension of his original DUKAT (Diagnose Und Kontrolle of Aethina Tumida, the small hive beetle) network project. During these visits, Neumann and his team conducted numerous investigations in collaboration with UWS staff on SHB biology, ecology and management. Australia was also used to provide field data complementary to that collected in the US and South Africa. This productive collaboration lead to a number of publications (Spiewok *et al.*, 2007; Buchholz *et al.*, 2008; Halcroft *et al.*, 2008; Neumann *et al.*, 2008; Spiewok *et al.*, 2008; Buchholz *et al.*, 2009; Greco *et al.*, 2009; Neumann *et al*, 2010; Halcroft *et al.*, 2011; Buchholz *et al.*, 2011; Spiewok and Neumann, 2012; Neumann *et al.*, 2012; Mustafa *et al.*, 2014). In November 2005, Neumann conducted a national seminar on SHB at UWS for researchers, government apiary officers and industry.

Discussion and conclusion

The initial discovery and apparent rapid spread of SHB in Australia provides some salutary lessons. Although a National Sentinel Hive Program was in place, SHB was not on a list of species to be monitored (in fact, no coleopteran was on the list).

There is genetic evidence that the introduction of SHB was from South Africa, and that it was the Durban strain (Anderson, 2002). There is considerable speculation as to how it arrived, but nothing definitive has been determined. However, it is clear that its initial arrival was not detected for some time and that movement of infested hives resulted in its spread, not only within the Sydney Basin and central coastal Qld, but subsequently into other states.

Australian beekeeping is highly migratory as beekeepers search for nectar flow, often crossing state boundaries in this quest, with hives and equipment. This migratory beekeeping has been exacerbated by loss of floral resources (Paton, 2008) and periods of drought. In addition, there is substantial hive migration for crop pollination. The largest pollination event in Australia is the almond crop, located in southern NSW, northern Vic and SA, with the second being lucerne (alfalfa) seed production in SA. It is been estimated that 164,000 hives are required during a season for almond pollination (Keogh *et al.*, 2010), and because almond crops flower during winter, strong hives need to be sourced from suitable locations; these are likely to be in environments that are favourable to SHB. Gordon *et al.* (2014) argued that the "nomadic" movement of hives in Australia creates potential for widespread disease in the honey bee industry, and reported that in the three months after completion of almond pollination in August (late winter) 2008, movement of hives occurred from Robinvale and Boundary Bend, on the Vic.-NSW border, to 49 locations ranging from south-east Qld to south-western Vic.

It also appears that the high density of feral *A. mellifera* colonies in many parts of Australia (Oldroyd *et al.*, 1997) and particularly within the Richmond district of the Sydney Basin contributed to the establishment of SHB in the local environment, making eradication unfeasible. The initial SHB survey, for example, reported that of 120 positive detections, 12 were in feral hives (Gillespie *et al.*, 2003).

As Australia is now the only major beekeeping country free from varroa mite, *Varroa destructor*, the SHB scenario provides a case study for the likely rapid spread of an exotic honey bee pest following its introduction, and confirms the importance of surveillance programs and early detection. Australia has an on-going National Bee Pest Surveillance Program (formerly the National Sentinel Hive Program) that was transferred from Animal Health Australia to Plant Health Australia in 2012. It has two major objectives: i. to facilitate the export of queen bees and packaged bees to countries sensitive to a range of bee pests and pest bees, by providing technical, evidence based, information to support Australia's pest free status claims during export negotiations and assisting exporters in meeting export certification requirements; and ii. to act as an early warning system to detect new incursions of exotic bee pests and pest bees (http://www.planthealthaustralia.com.au/national-programs/national-bee-pest-surveillance-program/).

In conclusion, SHB is now endemic to most mainland Australian states, although with the exception of NSW and Qld, has limited distribution. Its most severe activity is in coastal regions of NSW and Qld. It is unlikely to be endemic in locations not conducive to its survival, but it may cause major losses in areas not previously impacted when seasonal conditions provide a suitable environment (Annand, 2011). It remains to be seen what impact climate change will have on its distribution and abundance.

Concerningly, there is recent evidence from our and others' observations (Cheers, 2007, cited Annand, in 2011) that strong as well as weaker colonies can be severely impacted, particularly under high pest pressure. Consequently, integrated management practices will need to be diligently employed by beekeepers located in areas where SHB is endemic, and beekeepers in areas suitable for SHB but not yet severely impacted will have to remain vigilant.

Acknowledgements

We gratefully acknowledge Michael Franklin (UWS) for contributing the SHB distribution map. Our sincere thanks to the state apiary officers for valuable information on the status of SHB in their respective states: Hamish Lamb, Inspector (Biosecurity) Animal Biosecurity and Welfare, Queensland Department of Agriculture Fisheries and Forestry; Joe Riordan, Senior Apiary Officer, Department of Economic Development, Jobs, Transport and Resources, Victoria; Vicki Simlesa, Technical Officer Crocodiles & Apiary Officer, Northern Territory Government; Michael Stedman, Senior Apiary Inspector, Primary Industries & Resources South Australia; Bill Trend, Senior Apiculturist, Department of Agriculture and Food, Western Australia; Karla Williams, Program Specialist – Apiary, Department of Primary Industries, Parks, Water and Environment Quarantine, Tasmania. Also our thanks go to the many beekeepers that provided us with information regarding their own experiences with SHB.

References

Anderson, D. (2002). Research needs for the small hive beetle in Australia. *Honeybee News, November/December 2002, 10–11.*

Andrewartha, R. (2013). *Tasmanian plant biosecurity. Routine Import Risk Analysis (IRA) Importation of queen bees.* Department of Primary Industries, Parks, Water and Environment; Tasmania. http://dpipwe.tas.gov.au/Documents/IRA_Queen_Bees.pdf

Annand, N. (2008). *Small hive beetle management options.* Prime Fact 764, NSW DPI https://www.dpi.nsw.gov.au/data/assets/pdf_file/0010/220240/small-hive-beetle-management-options.pdf

Annand, N. (2011a). Investigations on small beetle biology to develop better control options. MSc thesis, University of Western Sydney, Penrith NSW, Australia. http://researchdirect.uws.edu.au/islandora/object/uws%3A11253/datastream/PDF/view

Annand, N. (2011b). *Small hive beetle biology: providing control options.* PRJ-000510 Pub. No. 11/044 RIRDC; Canberra, Australia.
https://rirdc.infoservices.com.au/downloads/11-044

Anon. (2003). *Small hive beetle national management plan.* Animal Health Australia. http://honeybee.org.au/pdf/SHB_FINAL_Mgt_Plan_AHA%20_31-10-03_.pdf

Bell, I. (undated). *Health Regulation Sub-program. Suspected Detection of an Exotic pest of Apiaries. Small hive beetle (Aethina tumida).*

Bernier, M., Fournier, V., Eccles, L. & Giovenazzo, P. (2014). Control of *Aethina tumida* (Coleoptera: Nititulidae) using in-hive traps. *Canadian Entomologist, 147,* 97-108.

Buchholz, S. B., Schäfer, M. O., Spiewok, S., Pettis, J. S., Duncan, M., Ritter, W., Spooner-Hart, R. & Neumann, P. (2008). Alternative food sources of *Aethina tumida* (Coleoptera: Nitidulidae). *Journal of Apicultural Research, 47(3),* 202-209.
http://dx.doi.org/10.3896/IBRA.1.47.3.08

Buchholz, S., Merkel, K., Spiewok, S., Pettis, J. S., Duncan, M., Spooner-Hart, R., Ritter, W. & Neumann, P. (2009). Alternative control of *Aethina tumida* Murray (Coleoptera: Nitidulidae) with lime and diatomaceous earth. *Apidologie, 40,* 535-548.

Buchholz, S., Merkel, K., Spiewok, S., Imdorf, A., Pettis, J., Westervelt, D., Ritter, W., Duncan, M., Rosenkranz, P., Spooner-Hart, R. & Neumann, P. (2011). Organic acids and thymol: unsuitable alternative control of *Aethina tumida* Murray (Coleoptera: Nitidulidae). *Apidologie, 42,* 349-363.

Fogarty, R. (2002). *Small hive beetle (Aethina tumida) in honey bees, Richmond NSW.* State Disease Control Headquarters SITREP No 1 Monday 28 October 2002, NSW Agriculture; Orange NSW, Australia.

Gillespie, P., Staples, J., King, C., Fletcher, M. J. & Dominiak, B. C. (2003). Small hive beetle, *Aethina tumida* (Murray) (Coleoptera; Nitidulidae) in New South Wales. *General and Applied Entomology, 32,* 5-7.

Gordon, R., Bresolin-Schott, N. & East, I. J. (2014). Nomadic beekeeper movements create the potential for widespread disease in the honey bee industry. *Australian Veterinary Journal*, 92, 283-290.

Greco, M., Hoffmann, D., Dollin, A., Duncan, M., Spooner-Hart, R. & Neumann, P. (2009). The alternative Pharaoh Approach: Stingless bees mummify beetle parasites alive. *Naturwissenschaften*, 97, 319-323.

Halcroft, M. T., Dollin, A., Francoy, T. M., King, J. E., Riegler, M., Haigh, A. M. & Spooner-Hart, R. N. (2015). Delimiting the species within the genus *Austroplebeia*, an Australian stingless bee, using multiple methodologies. *Apidologie* (submitted).

Halcroft, M., Spooner-Hart, R. & Dollin, L. A. (2013) Australian stingless bees. In *Pot-honey: a legacy of stingless bees, P. Vit, S. Pedro, D. W. Roubik (Eds)*. Springer; New York, USA. pp. 35-72.

Halcroft, M., Spooner-Hart, R. & Neumann, P. (2011). Behavioural defence strategies of the stingless bee, *Austroplebeia australis*, against the small hive beetle, *Aethina tumida*. *Insectes Sociaux*, 58, 245-253.

Halcroft, M., Spooner-Hart, R. & Neumann, P. (2008). A non-invasive and non-destructive method for observing in-hive behaviour of the Australian stingless bee, *Austroplebeia australis*. *Journal of Apicultural Research*, 47 (1), 82–83. http://dx.doi.org/10.1080/00218839.2008.11101428

Halcroft, M., Spooner-Hart, R. & Neumann, P. (2011). Behavioural defence strategies of the stingless bee, *Austroplebeia australis*, against the small hive beetle, *Aethina tumida*. *Insectes Sociaux*, 58, 245-253.

Heenan, L. & Heenen, P. (2007). Small hive beetle tray. *Honeybee News, March/April 2007*, p.33.

Keogh, R., Mullins, I. & Robinson, A. (2010). *Pollination aware case study: Almond*. RIRDC; Canberra, Australia. 8 pp. https://rirdc.infoservices.com.au/items/10-108

Keshlaf, M. & Spooner-Hart, R. (2013). Evaluation of anti-varroa bottom boards to control small hive beetle (*Aethina tumida*). *World Academy of Science, Engineering and Technology, International Science Index* 84, *International Journal of Biological, Food, Veterinary and Agricultural Engineering*, 7, 809-811.

Leemon, D. & McMahon, J. (2009). *Feasibility study into in-hive fungal bio-control of small hive beetle*. PRJ-000037, Pub. No. 09/090 RIRDC, Canberra, Australia. https://rirdc.infoservices.com.au/items/09-090

Leemon, D. (2012). *In-hive fungal biocontrol of small hive beetle*. PRJ-004150 Pub.No. 12/012 RIRDC, Canberra, Australia. https://rirdc.infoservices.com.au/items/12-012.

Levot, G. W. & Haque, N. M. (2006). Disinfestation of small hive beetle *Aethina tumida* Murray (Coleoptera: Nitidulidae) infested stored honey comb by phosphine fumigation. *General and Applied Entomology*, 35, 43-44.

Levot, G. (2007). *Insecticidal control of small hive beetle: Developing a ready-to-use product.* DAN216A Pt II, Pub. No. 07/146, https://rirdc.infoservices.com.au/items/07-146

Levot, G. W. (2008). An insecticidal refuge trap to control adult small hive beetle, *Aethina tumida* Murray (Coleoptera: Nitidulidae) in honey bee colonies. *Journal of Apicultural Research,* 47(3), 222-228. http://dx.doi.org/10.3896/IBRA.1.47.3.11

Levot, G. (2009). *Progress in developing strategies for the insecticidal control of small hive beetles.* RIRDC, Canberra, Australia https://rirdc.infoservices.com.au/items/09-182 2pp.

Levot, G. (2012). *Commercialisation of the small hive beetle harbourage device.* PRJ-004606, Pub. No. 11/122 RIRDC, Canberra, Australia. https://rirdc.infoservices.com.au/items/11-122.

Levot, G. (2014). *APITHOR™ small hive beetle harbourage trap safety and residue trials.* PRJ-008774, Pub. No. 13/106 RIRDC, Canberra, Australia. https://rirdc.infoservices.com.au/items/13-106.

Lounsberry, Z., Spiewok, S., Pernal, S. F., Sonstegard, T. S., Hood, W. M., Pettis, J., Neumann, P. & Evans, J. D. (2010). Worldwide diaspora of *Aethina tumida* (Coleoptera: Nitidulidae), a nest parasite of honey bees. *Annals of the Entomological Society of America,* 103, 671-677.

Malfroy, F. (2003). Letter to Editor. Response to "Discovery of the small hive beetle in A u s t r a l i a". *Honeybee News,* January/February 2003, p. 38.

Mustafa, S. G., Spiewok, S., Duncan, M., Spooner-Hart, R. & Rosenkranz, P. (2014). Susceptibility of small honey bee colonies to invasion by the small hive beetle, *Aethina tumida* (Coleoptera, Nitidulidae). *Journal of Applied Entomology,* 138(7), 547-550. http://dx.doi.org/10.1111/jen.12111

Neumann, P. & Ellis, J. D. (2008). The small hive beetle (*Aethina tumida* Murray, Coleoptera: Nitidulidae): distribution, biology and control of an invasive species. *Journal of Apicultural Research,* 47(3), 181-183. http://dx.doi.org/10.3896/IBRA.1.47.3.01

Neumann, P. & Elzen, P. (2004). The biology of the small hive beetle (*Aethina tumida,* Coleoptera: Nitidulidae): Gaps in our knowledge of an invasive species. *Apidologie,* 35, 229-247.

Neumann, P; Hoffmann, D.; Duncan, M.; Spooner-Hart, R. (2010) High and rapid infestation of isolated commercial honey bee colonies with small hive beetles in Australia. *Journal of Apicultural Research,* 49(4), 343–344. http://dx.doi.org/10.3896/IBRA.1.49.4.10

Neumann, P., Hoffmann, D., Duncan, M., Spooner-Hart, R. & Pettis, J. S. (2012). Long-range dispersal of small hive beetles. *Journal of Apicultural Research,* 51(2), 214-215. http://dx.doi.org/10.3896/IBRA.1.51.2.11

Neumann, P. & Hoffmann, D. (2008). Small hive beetle diagnosis and control in naturally infested honey bee colonies using bottom board traps and CheckMite+ strips. *Journal of Pest Science*, 81, 43-48.

Oldroyd, B. P., Thexton, E. G., Lawler, S. H. & Crozier, R. H. (1997). Population demography of Australian feral bees (*Apis mellifera*). *Oecologia*, 111, 381-387.

Paton, D. C. (2008). Securing long-term floral resources for the honey bee industry. UA-66A Pub. No. 08/087. RIRDC, Canberra, Australia.
https://rirdc.infoservices.com.au/downloads/08-087

Rhodes, J. & Mccorkell, B. (2007). Small hive beetle in NSW Apiaries 2002-6: Survey results 2006. *Honeybee News*, September 2007, pp. 27–28.

Spence, S. (2002). *Briefing Note: Summary of small hive beetle outbreak*. Technical Specialist (Farm Product Integrity), NSW Agriculture.

Spiewok, S., Duncan, M., Spooner-Hart, R., Pettis, J., Neumann, P. (2008). Small hive beetle, *Aethina tumida*, populations II: Dispersal of small hive beetles. *Apidologie*, 39, 683-693.

Spiewok, S. & Neumann, P. (2012). Sex ratio and dispersal of small hive beetles. *Journal of Apicultural Research*, 51 (2), 216-217. http://dx.doi.org/10.3896/IBRA.1.51.2.12

Spiewok, S., Pettis, J. S., Duncan, M., Spooner-Hart, R., Westervelt, D. & Neumann, P. (2007). Small hive beetle, *Aethina tumida*, populations I: Infestation levels of honey bee colonies, apiaries and regions. *Apidologie*, 38, 595-605.

Toffolon, R. (2002). *Small hive beetle (*Aethina tumida*) in honey bees, Richmond NSW*. SDCHQ SITREP No. 5, Friday 1 November 2002, NSW Agriculture.

Somerville, D. (2003). *Study of the small hive beetle in the USA*. A report for the Rural Industries Research and Development Corporation DAN-213A, Pub. No. 03/050 RIRDC, Canberra, Australia. https://rirdc.infoservices.com.au/items/03-050

Spooner-Hart, R. (2008). *Sustainable control of small hive beetle through targeting in-ground stages*. UWS-22A, Pub. No. 08/115 RIRDC, Canberra, Australia. https://rirdc.infoservices.com.au/items/08-115

Spooner-Hart, R. (2010). *Evaluation of anti-varroa boards to increase honey production in Australian honey bees*. PRJ-00355 Pub. No. 10/011 RIRDC, Canberra, Australia. https://rirdc.infoservices.com.au/items/10-011

Wade, R. (2012). Keeping out small hive beetles. *Aussie Bee*. http://www.aussiebee.com.au/abol-018.html

Robert Spooner-Hart[1,3], Nicholas Annand[2], Michael Duncan[1]

[1] School of Science and Health, University of Western Sydney, Penrith NSW 2750 Australia
E-mail: r.spooner-hart@uws.edu.au
[2] NSW Department of Primary Industries, Bathurst NSW 2795 Australia
[3] Hawkesbury Institute for the Environment, University of Western Sydney, Penrith NSW 2750 Australia

Appendix: recommendations of the COLOSS Workshop.

1. **Strategy:** Italian, European and global concerted efforts are urgently required to achieve standardised approaches to mitigate SHB impact.

2. **Research tools:** Specific, sensitive and feasible diagnostics need to be refined to enable reliable and timely use. Gaps in our knowledge (R&D resistance and resilience) need to be researched.

3. **Awareness:** Raising appropriate awareness and training amongst stakeholders and policy makers, including reporting of suspicious SHB finds and adequate compensation.

4. **Adaptive shift:** from eradication to control and management when SHB is established.

5. **Cooperation:** maintain beekeeper trust to ensure success of any strategy.

The Small Hive Beetle

Aethina tumida adult dorsal view (left) and ventral view (right). Photos © Per Kryger.

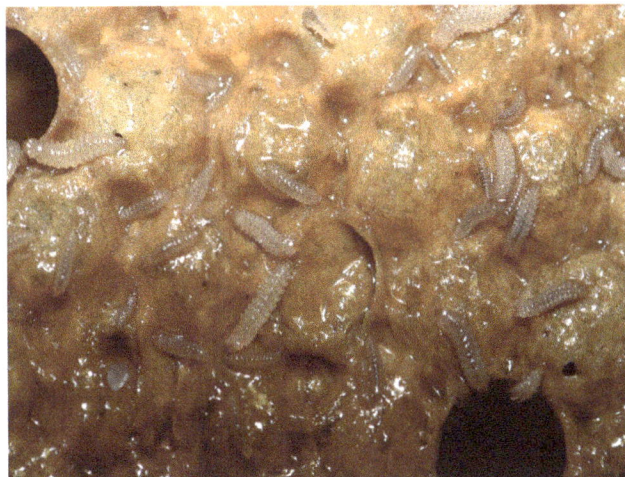

Aethina tumida eggs (left) and larvae (right). Photos © Marc Schäfer.

Aethina tumida wandering larva (left) and pupa (right). Photos © Marc Schäfer.

Contributors

Norman L Carreck

Norman has been keeping bees for thirty seven years, and has been a bee research scientist for twenty six. He has lectured about bees on all continents where bees are kept, has written many scientific papers, book chapters, conference contributions and popular articles, has edited several books and regularly appeared in the media in many countries. He is Science Director of the International Bee Research Association, Senior Editor of the *Journal of Apicultural Research*, and is based at the University of Sussex, UK.

Dr Franco Mutinelli

Franco has a degree in veterinary medicine from Bologna University, Italy, and holds the Diplomate of the European College of Veterinary Pathologists, and the Executive Master for management of health authorities from Bocconi University, Milan. Since 1989, he has been Veterinary Manager, IZS delle Venezie, Legnaro (Padova), Italy. He is Head of Experimental Veterinary Sciences Division, Head of Diagnostic Services Histopathology and Parasitology Department, and since 2003 Head of the National Reference Laboratory for beekeeping. His main field of activity is the diagnosis and control of honey bee diseases, environment monitoring, legislation, and education and training in apiculture, histopathology of animal diseases, neoplastic pathology and TSEs, rabies diagnosis, surveillance and control, laboratory animal husbandry and

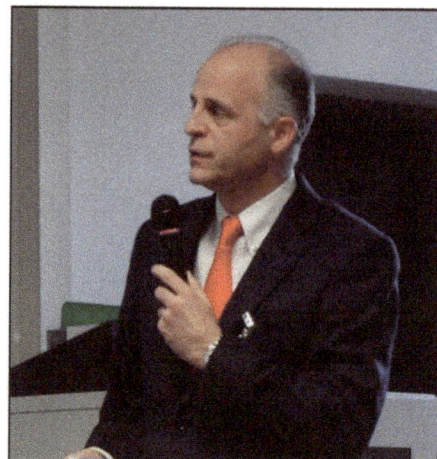

Dr Marie-Pierre Chauzat

Marie-Pierre graduated in Biology and Ecology, and has studied the ecotoxicology of the honey bee in the Unit of Honey Bee Pathology, Anses Sophia Antipolis (France) since 2002. The laboratory of Anses Sophia Antipolis is the OIE Reference Laboratory for bee diseases. After years of experience in bee pathology, Marie-Pierre gained extensive knowledge and experience in bee disease control and diagnostics, and, specifically, in field surveys. She is a member of the 'honey bee group of the EU Commission for the importation of bee hive products' (DG SANCO), and a member of the scientific committee of ITSAP, the French scientific and technical apiculture institute. She is a Management Committee member in the EU COST action SUPER-B that is currently working on sustainable pollination in Europe. Marie-Pierre has been Deputy Head of the European Reference Laboratory for honey bee health since April 2011. She has also been in charge of the EPILOBEE survey (A pan-European epidemiological study on honey bee colony losses) from the coordination of the field work (2012 -2014) to data statistical analysis.

Prof. Peter Neumann

Peter obtained both his MSc and PhD degrees in honey bee genetics under the supervision of Prof. Robin Moritz in Germany. He then carried on with a research focus on Cape honey bees and small hive beetles as a postdoctoral fellow at Rhodes University, South Africa with Dr Randall Hepburn. In 2006, he joined the Swiss Bee Research Centre in Bern, Switzerland as full time researcher, becoming head of the Pest and Pathogens Division three years later. In 2013, he established the Institute of Bee Health at the University of Bern and became Foundation Vinetum Professor of Bee Health at the Vetsuisse Faculty. His research now includes all aspects of bee health, with a strong focus on honey bee pathology including varroa and associated viruses. Peter is President of the global COLOSS (prevention of honey bee COlony LOSSes) association with more than 850 members from 96 countries.

Dr Jeff S Pettis

As a research entomologist in the USDA-ARS Bee Research Laboratory in Beltsville, Maryland, USA, Jeff leads a broad research effort to improve colony health by limiting the impact of pests and diseases on honey bee colonies. His research areas include; IPM techniques to reduce the impacts of parasitic mites and disease, effects of pesticides and pathogens on queen health and longevity, host-parasite relationships and bee behaviour. Jeff serves on several international committees concerning bee health and is frequently interviewed by the media for his opinions on worldwide pollinator declines. Jeff received an undergraduate and MS degree from the University of Georgia and his doctoral degree in entomology from Texas A&M University in 1992.

Prof. Christian Pirk

Christian has a full professorship at the University of Pretoria (UP), South Africa, and heads the vibrant Social Insects Research Group (UP). After finishing his studies in biology and mathematics at the Technical University, Berlin, Germany, he worked on reproductive conflicts in the Cape honey bee for his PhD under the supervision of Prof. Hepburn at Rhodes University, Grahamstown, South Africa. Some of his research interests include questions related to reproductive division of labour in social insects - how potential conflicts are solved. Chemical ecology - the organisation of groups, mechanisms of coordination and task allocation and the role and means of pheromonal communication modulating behaviour and its role in achieving coherent collective behaviour in social insects. He also works on honey bee diseases and the interactions and co-evolution between host and parasites, e.g. between honey bees and varroa or honey bees and small hive beetle.

Dr Robert Spooner-Hart

Robert is Associate Professor of Sustainable Plant Production Systems at the University of Western Sydney, Australia and former Director of the Centres for Horticulture & Plant Science and Plant & Food Science. He has over 35 years' experience in entomological research and development, with a major emphasis on sustainable pest management and beneficial insects in horticultural crops. He was leader of the Australian Eastern States Bee Breeding Scheme in the 1990s, and has supervised numerous PhD and MSc students in apiculture and meliponiculture. Following the discovery of SHB in Australia, he conducted a number of national projects on small hive beetle biology and management, and led the Australian team which collaborated with Dr Peter Neumann and his group. He has published two books, three book chapters and 90 refereed papers, including 12 on SHB, as well as numerous government reports, industry papers and consultancy reports.

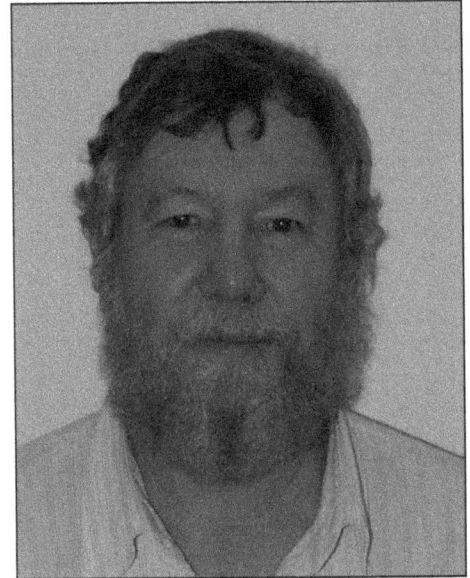

www.ingramcontent.com/pod-product-compliance
Lightning Source LLC
Chambersburg PA
CBHW050038220326
41599CB00040B/7199